INTERACTIVE TELEVISION *and* INSTRUCTION

INTERACTIVE TELEVISION *and* INSTRUCTION

A GUIDE TO TECHNOLOGY, TECHNIQUE, FACILITIES DESIGN, AND CLASSROOM MANAGEMENT

ROBERT H. LOCHTE
MURRAY STATE UNIVERSITY

EDUCATIONAL TECHNOLOGY PUBLICATIONS
ENGLEWOOD CLIFFS, NEW JERSEY 07632

Library of Congress Cataloging-in-Publication Data

Lochte, Robert H.
Interactive television and instruction : a guide to technology. technique, facilities design, and classroom management / Robert H. Lochte.
p. cm.
Includes bibliographical references and index.
ISBN 0-87778-252-0
1. Television in education. 2. Interactive television.
I. Title.
LB1044.7.L63 1993
371.3'358--dc20 92-31350
CIP

Printed in the United States of America.

Library of Congress Catalog Card Number: 92–31350.

International Standard Book Number: 0–87778–252–0.

First Printing: January, 1993.

TABLE OF CONTENTS

AUTHOR'S PREFACE

This book is written for instructors at secondary schools, colleges, and universities who will be using video conference equipment to offer instruction to students at more than one site simultaneously. Although I write primarily about an interactive television system where all people can both hear and see all other people involved, much of the information and method contained here will help instructors in one-way video, two-way audio systems, or other variations of the technology.

The background material for this book comes from three general sources. The first is the knowledge of audio and video production technique and technology I have acquired in twenty-two years as a working professional in the radio and television industry, and in the past four years as a member of the faculty in the Department of Journalism and Radio-Television at Murray State University. The second source is the preparation and original research I have done over the past two years to conduct workshops for interactive television instructors. This activity has forced me to reexamine my presuppositions about television and to begin to look at the medium as a technology that people can use, not vice versa. The final source for this manuscript has been the growing body of literature in the general area of distance education, about interactive media. Many citations are included in the bibliography.

THE HUMAN VARIABLES

In general, I feel that there are two qualities that separate the good interactive television instructors from the rest. The good instructors are prepared well in advance of the day and time they step into the classroom. The secondary school teachers who have participated in my workshops have, on the whole, adapted to the technology more readily than university faculty — not because they are better teachers, but because they are used to teaching the same subject five days a week with five or more classes back-to-back. Their daily routine requires more attention to the logistics of classroom management and advance preparation.

The other determinant of a good interactive television teacher is consideration for the students. Taking the time to plan interaction with the students as a regular part of your classroom routine is important because it provides you with feedback on the learning process, and it tells them that you care about them and want to involve them in that process.

Thinking about the students will also improve your management of logistics for distant classrooms and suggest contingency plans in case the technology fails.

Both preparation and consideration are human variables which you can control. In fact, they are both important elements to instructional method for any classroom environment. Many instructors with whom I have worked tell me that teaching with interactive television has made them better teachers overall because they have had to reconsider the basics of teaching and learning. Perhaps this is the best way to make the technology more transparent for the people involved.

MY OBJECTIVES

The purpose of this volume is to provide a companion piece for workshops where teachers will get hands-on training with interactive television equipment in an actual classroom.

Secondarily, it is meant to be a reference for teachers, administrators, aides, and perhaps students involved in classes offered via interactive television. It is not intended to be a comprehensive text on the subject. You simply cannot master this, or any other, instructional method by reading a book.

INTRODUCTION TO TECHNOLOGY

The first three chapters describe the basic technology of television production and distribution, from both historical and practical perspectives. It is important that you learn the "grammar" of this medium, and I have attempted to describe it in nontechnical language where possible. Bear in mind that electronic communication was our first high-tech industry. From the moment of Morse's first telegraph transmission, communication was freed from the limitations of transportation. As an institution in modern society, education has been slow to adopt this technological development and is now attempting to catch up. I do not suggest that you latch onto every new gadget that comes down the pike. On the contrary, I think that technology only works as well as people want it to, and that we need to set human objectives of teaching and learning before we invest dollar one. Nevertheless, you need to know the functions and limitations of television, just as you have this knowledge about your automobile, to be able to use the medium effectively.

THE CLASSROOM EXPERIENCE

The next three chapters deal with the actual interactive television classes. Based on my experiences at Murray State and conversations with colleagues at other schools, I have identified two potential problem areas which require attention in the preparation and presentation stages. The first is the technology itself, which does not always work as well as we wish it would. The second is the low-tech activity of getting classroom materials to the distant sites, and handling assignments and tests. If you can formulate plans to deal with these, you will find that interactive television requires little behavior modification on your part.

In these chapters, I propose a model that suggests you do several things. You should consider an interactive television course to be a high-grade correspondence course rather than an extension of your regular classroom. In preparation, this means establishing a logistical timeline as well as a content outline for your class. You must also take a teamwork approach to the class. Members of the team will include your students, your classroom aides and technicians, and perhaps program administrators. Communicating with everyone is important. So is achieving a level of comfort in the classroom environment so that the interactive televison technology becomes second nature, and the focus of activity is on learning.

As an integral part of any interactive television class, I suggest an ongoing practical evaluation that provides a feedback loop — a communication mechanism for teachers and students to make decision-makers aware of progress and problems. The other facet to this evaluation structure is that decision-makers plan in advance how to use this information to improve the project at any point in time when improvement or change is indicated, not as a summative set of recommendations at the end of a school year. Although the structure of this evaluation process can be as complex as you wish, it really boils down to asking, as former New York City Mayor Ed Koch used to, "How am I doing?" of everyone involved on a regular basis. By creating this mechanism at the outset, and using the information it generates, you should be able to solve problems both technical and nontechnical before they become serious enough to derail the project, as well as fine-tune the technology to promote learning and fit your teaching style.

OTHER USEFUL INFORMATION

The final two chapters are *lagniappe*, as my Cajun cousins say. I describe how to set up your own instructor workshop using this volume as a resource. Many readers will need to do this for in-service training or for orientation of new personnel. Using existing faculty is more cost-efficient and frequently more effective than importing experts for such an event. My emphasis in these workshops is not only to get hands-on experience as a teacher, but also to get the feel of being a student in the interactive television environment.

Finally, I ask you to think about the physical and technical components of the interactive television classroom. Perhaps you should put off reading the final chapter until after you have taught a class via the medium. Once you buy into any distance learning technology, you quickly learn that someone is always creating a new gadget to improve your life and that you can spend as much money as you have and still not be state of the art. Back to my original premise that technology should be the slave rather than the master of us humans, I hope this last chapter will encourage you to design learning environments for your students, then ask for the technical assistance to make such environments possible.

ACKNOWLEDGMENTS

Several people have helped me with this project and I am indebted to them all. First and foremost is Dean Viola Miller of the Division of Continuing Education and Academic Outreach at Murray State University, who hired me to help get our interactive television network off the ground. From her staff, Mary Boaz, John Hart, Laura Wells, John Yates, and Bonita Hale have all been most cooperative. Orville Herndon has provided layout and illustration assistance, without which I could never have finished this manuscript.

I want to thank my boss, Bob McGaughey, and my Dean, Gary Hunt, for allowing me release time to work on the interactive television network, and the participants at my workshops over the past two years who have provided feedback on this text. Also my students in Corporate and Instructional Video have provided me a forum for some of the ideas contained here.

Finally, I appreciate the folks who took the time to read this manuscript at various stages and tell me honestly what they think about it. These include Doug Nesbit of Murray State's Faculty Resources Center, Bob Swisher at the University of Oklahoma, and Judy Euller at the University of North Dakota.

Robert Henry Lochte
June, 1992

INTERACTIVE TELEVISION *and* INSTRUCTION

CHAPTER 1

WHAT INTERACTIVE TELEVISION IS

To this point in time, most of us have thought of television as a familiar source of information, entertainment, and advertising that we consume rather passively. In four decades or so, television has emerged as a fixture in our daily environment. There are, however, new concepts of television that may change forever the way we think of the medium. Consider the differences between broadcast or cable television and a program you rent from your local video store, or one you create with a camcorder, or the experience of playing a video game. In each of these instances, you are actively choosing how you will use television and creating your own media environment. You are no longer a purely passive consumer, but have in fact become a program director.

The term interactive implies some form of two-way human communication. Many of these new television environments, including the subject of this book, will entail interactivity among people or with other environments (i.e., databases) that people have created. Yet the term Interactive Television (ITV) means many things to many people.

INTERACTIVE BROADCAST TELEVISION

One of the first television programs to encourage audience participation was Jack Barry's *Winky Dink*, a children's show from the early 1950's. At one point in the program, Winky Dink would draw a picture and ask the viewers to help him complete it. To do this, you needed an accessory, a clear cellophane overlay for the TV screen which was available at local toy stores, and some crayons. Then you could draw along with Winky Dink. There is no telling how many TV sets ended up with crayon marks permanently affixed to the picture tube for failure to use the cellophane overlay.

For many years, the broadcast and cable television industry has employed interactivity via telephone lines to encourage viewer response. To the telemarketer, interactive television means using an 800 or 900 phone number so viewers can order records, knives, or the latest gadget guaranteed to firm your tummy and shrink your thighs. Likewise, talk shows and some game shows can allow viewers to call in and participate in the program.

Another form of interactive television common in the broadcast and cable business is the use of live satellite transmission for news and sports events. It is normal to tune in *Nightline* or CNN or a football game and see personalities who can communicate with each other by both audio and video links. Although such productions are technically complex and expensive, you will be able to duplicate this type of television environment in your classroom at a reasonable cost.

COMPUTERS AND INTERACTIVE TELEVISION

Adding computers to a television system allows still another form of interaction by being able to transfer data and text between the parties and by being able to store information to be used at a more convenient time. Few of us who played PONG or PACMAN for the first time realized how much interest video games would generate. Communicating via computer bulletin boards has now become common, as has computer-assisted instruction for complex subjects

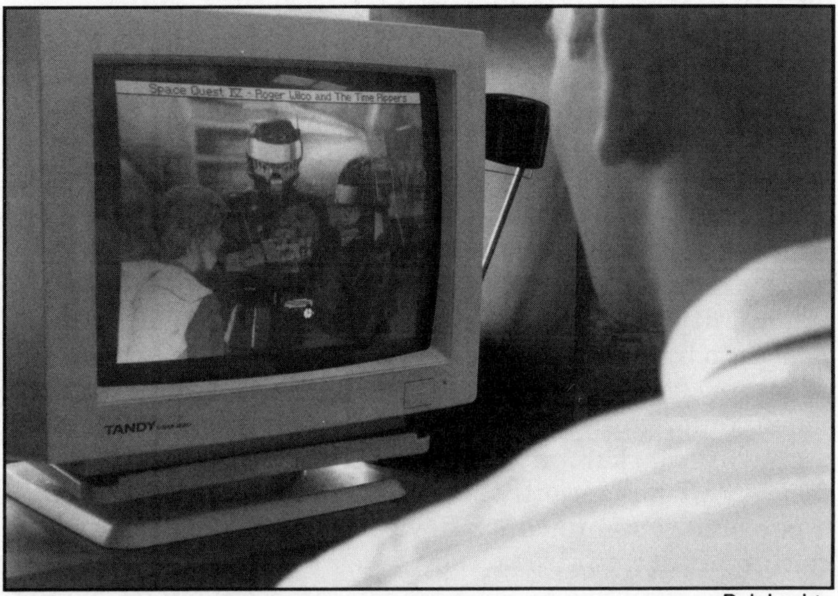

Bob Lochte

One form of interactive television, the videogame.

such as advanced mathematics and for more mundane learning tasks.

Using computers as control devices has opened up other possibilities for interactivity. From file management software came hypertext documents. By incorporating video from CD-ROM drives or laser disc players in place of text or data files, hypertext became hypermedia, a flexible and potentially powerful interactive instructional technology. Now desktop video production is possible as both original video and audio signals may be digitized and processed by computer with the same ease as desktop publishing software converts text and graphics to publication quality layouts. The next generation of computer-based interactive television will create environments so credible that they are called virtual realities.

TELECOMMUNICATIONS AND INTERACTIVE TELEVISION

At the New York World's Fair of 1964, AT&T introduced a form of interactive television—the Picture Phone. This device allowed telephone users to see as well as hear each other. Transmission could be accomplished by conventional coaxial cable, by microwave links, and later by communication satellites and fiber optics. The business community soon saw the advantages of holding face-to-face meetings by two-way television rather than paying travel expense. As costs came down, business teleconferences and video conferences became more common and now represent a growing market for the telecommunications industry. In fact, AT&T has recently begun to sell the Picture Phone as a personal communication device, more than a quarter century after its introduction.

There were two experiments in interactive television by cable systems in the 1970's that bring us closer to the present subject. In Columbus, Ohio, Warner-Amex Cable established the QUBE system, which allowed subscribers to interact with information provided on certain channels by using the digital touch-tone pad on their television remote control. Among the activities available were banking, shopping, and public opinion polling. Meanwhile, in Irvine, California, a professor from UCLA helped the local cable system design and build a two-way interactive audio and video channel for use by the local schools. And this is where we come in.

For the purposes of this book, we shall use this latter description of interactive television. There will be some variables depending on the transmission system chosen in your school. For example, if you are using a satellite transponder to reach many distant sites, economy will probably dictate buying only one uplink. Thus, you will have one-way video with two-way audio, most likely on a regular telephone line. Or your interactive response mechanism may be a digital pad connected to a computer database. If you are using dual channel microwave or dual channel broadband cable television, you will have real-time, full-motion video going in both directions, as well as two-way audio and computer communications. If you are using a digital network, where the television signal is compressed into a stream of bits and expanded on the other end, you will still be able to combine audio and video with text and data in both directions. Your picture, however, will be of inferior quality to what you are used to seeing on television. As a matter of fact, few of us will achieve network quality in our televised classes.

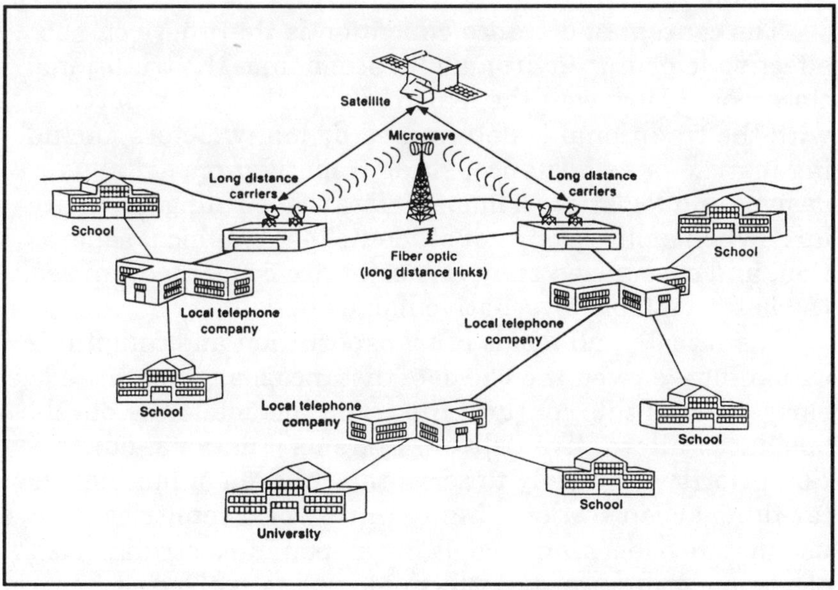

Office of Technology Assessment, 1989

A complex ITV distribution system involving satellite, fiber optics, and standard telephone lines, all switched by a local telephone company.

DISTANCE LEARNING AND INTERACTIVE TELEVISION

Nevertheless, all these means of transmitting interactive television work and can be useful. This text is designed to help you understand how to use the medium regardless of how the signal gets from one place to another. You will find that interactive television is superior to correspondence courses by mail, canned audio- or videotapes, or one-way educational television, if you become comfortable with this medium and use the flexibility it provides. It will probably never be a suitable substitute for a good teacher interacting in person with motivated students and should not be considered as such. The economics of education, however, will not always permit this optimal condition.

The concept of distance education is that there can be effective learning environments other than the traditional classroom. How well the learning experience compares with the traditional is determined by many factors, including instructor competence, student motivation, syllabus design, and logistics. Thus, in distance learning, particularly when small groups of students, geographical separation, and specialized subject matter are concerned, we seek the best available alternative means of delivery.

Historically, all forms of transportation and communication have served the cause of distance learning. Since the early years of this century, Australia has made the development of alternative delivery systems a national education priority, primarily to serve the children of families in far-flung sheep stations. Many third-world countries now use distance learning models developed in Australia. As electronic communication technology has evolved, it has supplemented and often replaced older distribution media. Now education at all levels is available in many countries at locations convenient to potential students. Interactive television is merely a new wrinkle in an old fabric covering a wide variety of educational experiences and opportunities.

However your course is organized and delivered, the quality of the learning experience depends largely on the ability of the human beings involved to cooperate with one another. As an instructor, this means communicating effectively with your students and your facilitators and managing the classroom environment. Developing these skills is as important as mastery of your subject matter. With interactive television there is a special challenge because we do not fully understand the nature of the electronic classroom environment we have created. At various times it resembles broadcast television, a telephone call, a NASA mission, a business teleconference, or a regular class. In order to teach well, you must focus even more diligently on the human needs of your students — what they are learning and how you can manage the class to optimize this learning experience.

CONCLUSION

One further note of caution. Let the technology work for you rather than modifying your behavior to suit the interactive television environment. This may not always be possible but will be achievable most of the time if you think your way through problems that arise. In general, if you are well organized, and plan more material for each lesson than you think you can cover, and make an effort to interact with your students by whatever means, then this should be an interesting and perhaps exciting experience. And, of course, if you find yourself stymied by the technology, you can always ask your students. They probably know more about it than you do anyway!

CHAPTER 2

HOW TELEVISION WORKS — VIDEO

With television, you can capture the sights, sounds, and motion of the world around you, convert them to a stream of electrons, channel them into a cable or cast them into the air, recapture the electrons, and change them back into sights, sounds, and motion. Sounds like magic, doesn't it? Perhaps it is. Nevertheless, television is the process of producing, transmitting, and reproducing audio and video images. It is important to remember that this process is not technically perfect and, with the opportunity for human intervention at each stage, is governed as often by Murphy's Law (anything that can go wrong, will!) as by the laws of physics. Compromise is the rule rather than the exception.

Transmitting the signal is usually not the concern of the classroom instructor. As mentioned in the last chapter, there are several ways to get the television signal from point A to point B, and they all work. If the distant class is not receiving your signal, there will be certain procedures for checking your transmission equipment, some as simple as seeing whether it is plugged in, that will be peculiar to your classroom setup. Hopefully, your technicians will have provided you with a written set of instructions and some basic training. A backup communication system, usually a telephone, is also helpful. We shall cover contingency plans in a later chapter.

The production and reproduction of the television signal are, however, the responsibilities of the classroom teacher, and these are the subjects of this and the subsequent chapters.

THE TELEVISION CAMERA

Think of an old man afflicted with myopia and emerging cataracts. He would see the world much as a television camera does. He would need sophisticated lenses to bring his field of view into focus, would perceive distorted horizontal distances, and would require additional light to be able to see at all. Paradoxically, too much bright light would interfere with his ability to see. Compared to the normal human eye or to 35mm film, the television camera produces a much lower resolution picture, or one that is not as sharply in focus. The camera contains a pickup device, sometimes a tube but most often a CCD chip. The lens focuses the image on this device, and an electron beam scans it in horizontal lines, 525 times per frame, at a rate of 30 frames per second. In chip cameras, light sensitive pixels on the CCD generate these lines sequentially without an electron gun scanning it. High Definition Television (HDTV) systems, which approximate the resolution of 35mm film, scan more than twice as many lines per frame. Low resolution presents a problem at creating good television pictures of closely patterned, repetitive graphic designs, like houndstooth checks in clothing, because subject lines correspond with picture scanning lines. These will set off a visual echo effect, called *moiré* which appears as swirls of colors emanating from the picture.

So the quality of the lens is critical to achieving a usable television picture. Cameras will come with either a fixed focus, usually wide-angle, lens or one with variable focal length, commonly called a zoom lens. The zoom lens allows more flexibility in camera placement. You can, for example, place the camera at the back of the classroom, atop the monitor that shows your distant class, so when

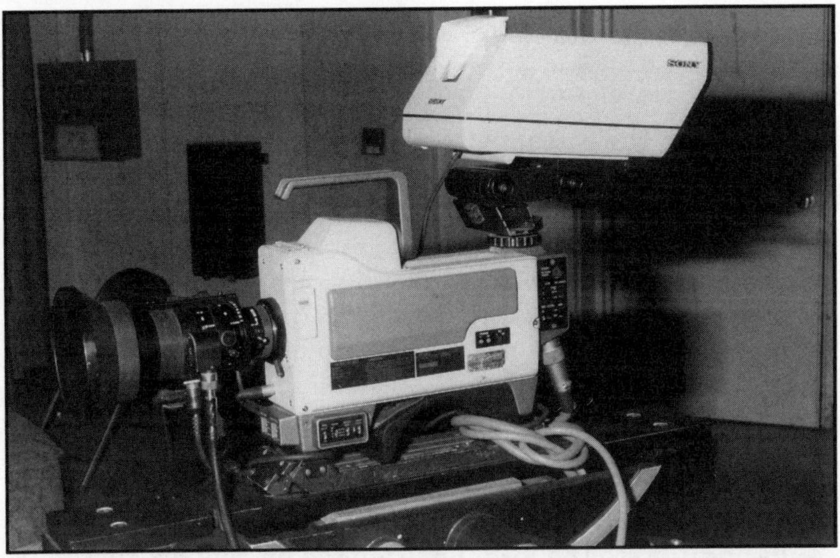

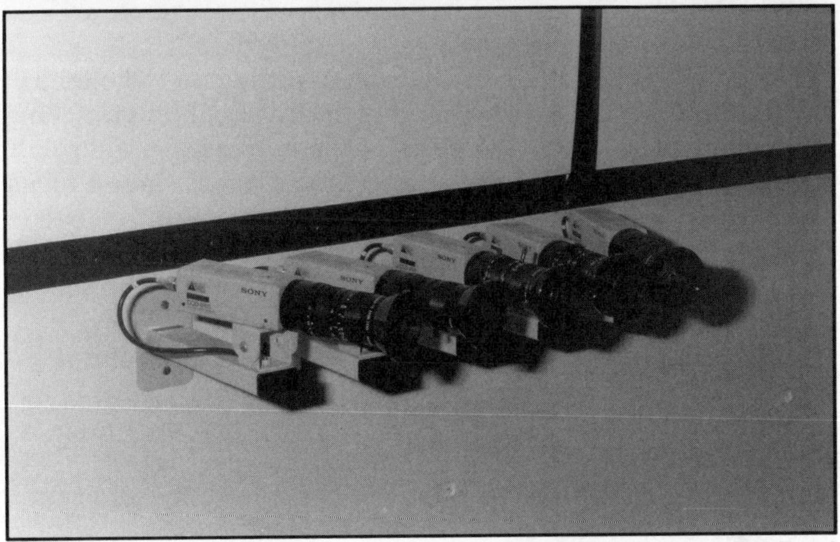

Murray State University, Rod Freed

A typical television studio camera with zoom lens and large viewfinder and surveillance type television cameras adapted for ITV use.

you look at those students, you also look into the camera and have eye contact with them. Using a zoom lens, you can narrow the field of view so that you appear larger on the screen at the distant site and give the illusion that you are closer.

There are some compromises with a zoom lens, but none are serious enough to keep you from using this valuable device. When you zoom in, you lose depth of field, meaning less of the picture is in focus. Your face may be sharp and clear, but if you hold up an object in front of you, it will probably be soft and slightly blurred. You also distort perspective so that objects in front of or behind you appear closer than they really are. Finally, you lose effective light level, which can make the entire picture lose resolution and appear grainy.

In addition to the focal length or zoom control, these lenses also have focus and iris controls. To preset your focus, zoom in as far as you can, to the tightest closeup setting, and focus on the image you want to see. The lens will remain in focus on that image at any focal length. This is especially important for your overhead camera, where all graphics will remain in the same plane. If you cannot focus on this image with the tightest closeup, you need to move the camera farther away, or install a range extender (a magnifying glass that screws onto the lens), or a combination of both.

The iris controls the amount of light that enters the camera and strikes the pickup device. Normally, this is an automatic function. If you find, however, that the image is dark (or in the case of a zoom lens with a very tight closeup), you may need to open the iris manually. Another alternative is, of course, to add more light.

Some of your "friends" may tell you you look fat on television, and in fact, you do. Our television system has a built-in distortion that spreads out horizontal space. You will find that objects, or words on a page, appear to be farther apart than they really are. Many of your visual aids prepared for an overhead projector will need to be recomposed for television due to this problem.

GETTING A GOOD PICTURE

Most of the compromises in producing good television pictures involve lighting. Television cameras like high average brightness, which means if you want something to be seen, you have to put a light on it. Typically you will want to see an entire class of students, so the entire classroom must be illuminated. Fortunately, advances in CCD technology are such that television cameras can function in many existing classrooms.

Television cameras do not like high contrast ratios, which means that the difference in illumination in the brightest and darkest parts of the picture cannot be very great. The illumination source of choice would be even, soft light. Fluorescent or reflected incandescent lights will both work provided they furnish the minimum illumination level for the cameras you have chosen.

The simplest and most powerful way to control light levels is to vary the distance between the light and the subject. This is an inverse square relationship, meaning that if you cut this distance in half, you gain four times the amount of effective illumination. With light fixtures already in place, it is probably easier to move the people.

The camera actually sees reflected light, so the components of the picture and their reflective values are as important as the lighting. If you wear a navy blue blazer with a white blouse, you are creating a high contrast scene that the camera cannot handle. Change to a pastel blazer or a gray blouse, and the camera will like you much better. If the wall behind you is white, or considerably lighter than your face, the camera will silhouette you against the wall. We shall discuss these topics at some length further on.

You should also be aware that different light sources

have different color temperatures, or qualities. Incandescent light is red/orange; quartz light is yellow; and fluorescent light is blue, as is sunlight. The camera will most likely have filters to correct for the proper lighting. Use the outdoor filter for fluorescent and the indoor filter for all other types of lighting. Your colors will not be perfect, but they seldom are even under rigidly controlled professional conditions.

You need to shade or drape all windows in the television classroom. Sunlight coming from one angle will create dark, harsh shadows from the reverse angle, and give the cameras fits. It will also create color temperature problems.

TELEVISION MONITORS

Most of what you need to know about television monitors you already know or can guess. Look at one. What do you see? A small box, with a two-dimensional picture that is slightly wider than it is tall, from which spews generally inane programming that people watch anyway. Content criticism aside, think about the way you watch television. After 10 minutes or so, do your eyes start drifting around the room, focusing on everything but the television set? Perhaps this concept, a facet of the phenomenon known as persistence of vision, explains why there are breaks in the programming every 8-10 minutes.

How the television signal is reproduced and seen and heard is as important as how it was produced in the first place. Consider what you already know about the television picture. If it is to be seen on a small box, should you choose closeups or wide shots? Closeups of course. Many a wide-screen spectacular motion picture has lost impact on television for just this reason. Television miniaturizes the world. You must do as much as possible to compensate for this effect.

THE PICTURES YOU SEE

The term aspect ratio describes the ratio of the width to the height of a field of view. The aspect ratio of the television screen is 4x3, or slightly wider than it is tall. Knowing this should suggest two things about composing pictures for television. First, horizontal composition is better than vertical, so any vertical slides, charts, graphs, or other visual aids you have will probably have to be redone to fit on television. Second, you can use the zoom lens to isolate part of a vertical picture, such as a standing human being, and make it look better. Why do you think that television newscasters are typically framed "head and shoulders"?

Murray State University, Rod Freed

Two video monitors at the teacher's workstation. The one on the left shows the distant site; the one on the right, the transmission.

Establishing the illusion of three dimensions, or four if motion is considered, is a challenge. Artists, photographers, and filmmakers use variations of light and shadow, called subject modelling, to create 3-D effects. Modelling is difficult in television due to the technical restrictions that preclude high-contrast, or low-key, lighting design. There are some tricks you can play if you wish to experiment. Motion toward or away from the camera is probably the easiest. Another is to stage foreground or background action, using a wider lens angle and closer camera position to create more depth of field, or focus. Watch television programs critically for these techniques. You may find examples you can adapt easily to your presentations.

And those presentations will need visual variety. Remember that your distant students will only see what you show them on television. Every 8-10 minutes, they will begin to look around the room regardless of what you do. The key is to get them refocused as soon as possible. The longer the class period, the more variety you will need. In your traditional classroom, you can get up and move around, or redirect attention to the blackboard, or have students refer to printed material. All these activities may not work with your televised class, so you will need to develop others. The simplest is to change camera shots periodically, a function that can be automated, but there should always be a reason for these visual transitions.

Now here is something about television monitors that you probably did not know. They do not reproduce the picture material faithfully, partly due to transmission compromises and partly due to receiver design. Each time a television picture is transmitted, about 10% of the border area is lost.

In broadcast television, directors figure they will lose as much as one third of the picture by the time it gets to the home screen. Therefore, you must position all information you want your students to see near the center of the screen and leave some border area to protect yourself. Most picture tubes, which work in the reverse of camera pickup tubes, have even lower resolution than television cameras, usually 350 horizontal lines. Monitors larger than 15" diagonal lose effective resolution as the size of the screen increases. This is especially noticeable in wide screen projection systems. Even though you produce a well-focused picture, your students may not see it that way.

One last comment about monitors. Ambient room light can detract from the ability to see the picture well. It is difficult to position monitors in the interactive television classrooms so that they are isolated from this ambient light because you have to illuminate the room fully and evenly for the television cameras that are also present. Angling or shading the monitors may be necessary.

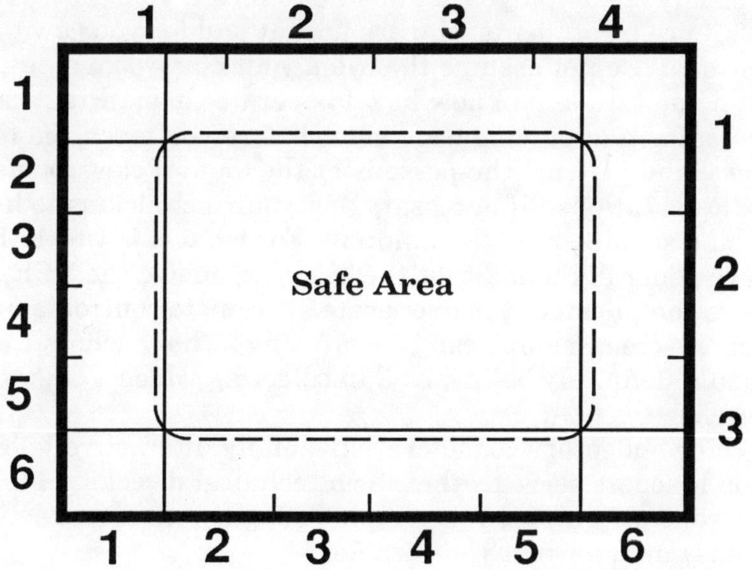

Orville Herndon

The area in the center of the screen is safe from transmission loss in the border areas. Plan to lose 10% of the picture each time it is transmitted.

OTHER VIDEO AND CONTROL EQUIPMENT

Recently, NBC did an experiment. The network mailed out 100 VHS tapes to VCR owners selected at random and asked them to record "The David Letterman Show" and send the tape back in the stamped mailer enclosed. Less than a third of the VCR owners could record this program correctly. Can you imagine what our highways would look like if drivers had the same level of competency operating their automobiles?

No one knows for sure why humans have so much trouble operating a VCR. More than two-thirds of American households have them, and it would seem we could master the device. The point is that you will probably have one or more VCRs in your interactive television classroom, typically one to playback and one to record your classes for reference by students who are absent and for your own critique. Do not assume that you, your classroom aides, or even your students know how to operate them. Practice recording and playing back video before you teach the first class, and instruct the persons at the distant classrooms in VCR operations. If necessary, get your technicians to help. If all else fails, read the manual! For what it is worth, I have generally found that the more expensive the VCR, the more complicated it is to operate. A remote control is nice, but on-screen menus can be confusing. The teachers should definitely be involved in selecting which VCRs to purchase.

For budgetary considerations, many interactive television teachers serve as their own technical directors. Most of the control equipment is simple enough so that this activity is easy and, for many, desirable.

Whenever there are multiple cameras, there is also a video switcher that allows you to choose which one to transmit. Other video sources, such as a VCR or computer, are connected to the same switcher so that you may choose them as well. Most instructor-controlled interactive televi-

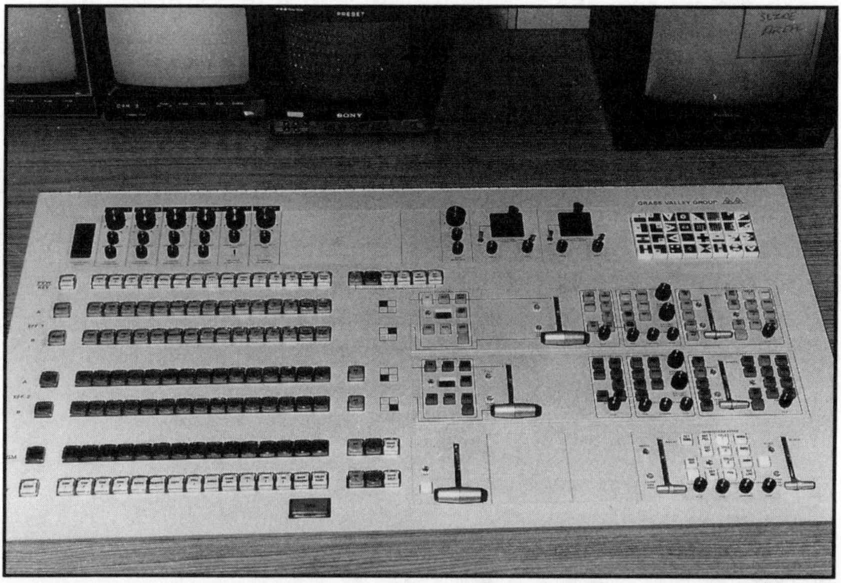

Murray State University, Rod Freed

A professional video switcher.

sion classrooms use a cuts-only switcher which instantly changes from one picture to another. Many of these also permit the instructor to select video sources at the remote sites. More sophisticated and costly video switchers include transitions such as dissolves, where one picture fades out as the next fades in, and wipes, where one picture pushes another off the screen. Inserting one picture into another, laying text over a picture, and combining two or more images are other effects available from video switchers. The more complicated the switcher, the more likely you are to need a separate technician to handle this duty. You may also want to think about preproducing and storing on videotape or laser disc any complex television material, especially resources you wish to use again.

Personal computers can have several useful functions. They can serve as video switchers. They can be a video or an audio source. They can supply text and graphics, and when set up in a network, can become interactive devices themselves. You can use bulletin board services and computer conferencing as a flexible seminar, permitting students to log on and join the discussion as their schedules permit. And you can deliver all your remote site examinations by computer, with your students returning their tests electronically into your files. I think it is best to set up a separate and parallel computer network to the interactive television system. You will find inexpensive uses for this

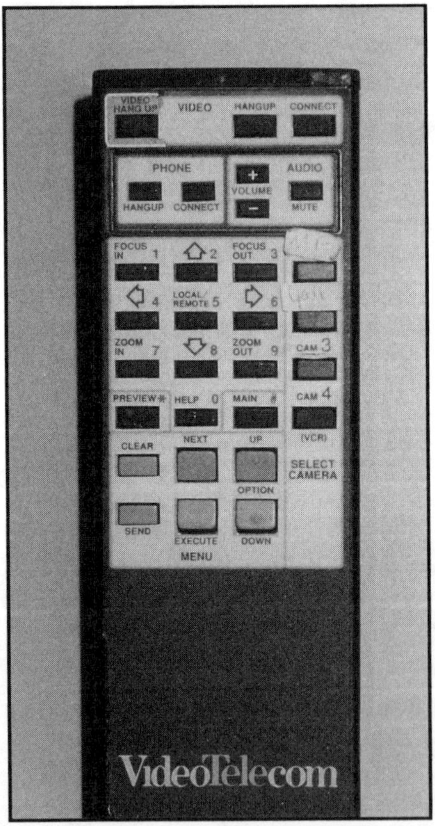

Murray State University, Bob Lochte

A simplified, remote-control, video switcher, designed for ITV utilization.

network that do not require live video. And if you are using a computer for video control, having a separate computer network minimizes the risk of crashing the entire system when you get stuck in a spreadsheet or word processing program. Plus, it provides an excellent backup to the interactive television network.

If your cameras have remote-control zoom, focus, and camera movement functions, you may also want to consider a technician. Most students will not talk very long, either making comments or asking questions. You need to concentrate on what they are saying, not on getting a decent picture. An alternative is to install more cameras with fixed focus on individuals or smaller areas of the classroom.

CONCLUSION

Practically, if you understand the limitations of the television picture and the setting in which it will be viewed, you will know enough to plan your classes initially. Above all, do not be afraid to experiment, and do not be afraid to ask your students how things look or how they could be improved. Soon they will begin to volunteer information, and that's what interaction is all about.

CHAPTER 3

HOW TELEVISION WORKS — AUDIO

In television production, audio is always more problematic than video. Think of an evening newscast. The televison station sends one composite audio and video signal into your home. Yet on the set, there are also separate audio systems for the director to communicate with the crew and for the producer to communicate with the anchors. The audio engineer is responsible for designing, operating, and maintaining three audio systems, while the video engineer has but one. These communication systems are especially complicated on live remote broadcasts.

Producing good quality audio is a challenge in itself. But the job is made more difficult in several ways when that audio is produced for television. If you were recording a singer, you would choose a microphone that faithfully captured or even enhanced her voice. If you put that singer on television, you must choose a microphone that looks good on camera, or figure some way to hide it from the camera. If you shoot with a telephoto lens to get a closeup shot, and record the sound with a microphone attached to the camera, the audio will sound distant and will confuse or annoy the audience because it does not match the video. Remember this the next time you get out the camcorder for a child's birthday party or a family picnic.

Most of us grew up watching television when audio was secondary to video mainly because television sets had poor sound systems. But television audio has never been inherently inferior. In fact, television can transmit the same dynamic range and frequency spectrum as FM radio. In the past decade, television set manufacturers have begun to exploit this advantage, and program producers have responded by recording shows in stereo with the same care given to audio-only recordings. And, believe me, poor audio will kill a television production regardless of the video quality.

As with video, an audio system captures the sounds of the world around us, changes or transduces them into an electrical signal that can easily be controlled, stored, and transmitted, and changes this signal back into sound. Rather than attempt to explain the complexities and craft of audio engineering, we shall look, as in the previous chapter, at the devices that capture, reproduce, and control the sound.

MICROPHONES

A microphone is a variety of transducer that contains a mechanism sensitive to sound pressure waves and an electronic network that causes those pressure waves to generate electricity. It simulates the mechanism of the human ear and the auditory nerves. Many microphones can "hear" better than most humans, hence audio engineering is further advanced than its video counterpart. In audio systems, the dynamic range (volume) and frequency response (pitch) characteristics of the transmission channel usually limit the quality of the output sound more than the microphones. Consider the differences in sound quality among a compact disc player, an AM radio station, and a telephone call.

Microphones are generally designed to pick up sound from all directions (omnidirectional) or to reject sound from one or more directions (directional). Most directional microphones can be aimed at desirable sources of sound and away from unwanted sources, so they are more useful in the interactive television classroom. It will, however, take more directional microphones to cover the room than

Murray State University, Bob Lochte

Voice-actuated microphones mounted on stands to isolate them from the table surface.

omnidirectional ones, so the expense is greater. Nevertheless, purchasing microphones is a poor place to scrimp on the equipment budget. The instructor will probably have a hard time seeing the distant students on television as well as he or she would like. Reducing the efficiency of audio communication can make the classroom unbearable.

The same inverse square law that applies to lighting applies to the relationship between microphone-to-source distance and volume. If you cut the distance between the microphone and the person speaking in half, the volume quadruples with no adjustment to any audio control equipment. This is the single most powerful volume control device you have available, and also the least expensive. Moving the microphone or the person is even better than having him or her speak up because decreasing the distance also reduces the probability that the microphone will capture unwanted ambient sound. In the best of all possible worlds, you would have one microphone for each person in the classroom, mounted as close as possible to each person's mouth.

There are both wired and wireless microphones. Wireless microphones, which utilize a radio transmitter and receiver, are more expensive than wired ones, are powered by batteries that need to be checked before each class, and require a line of sight between transmitter and receiver. If you clip the transmitter to your belt and turn so that your body is between it and the receiver, you will most likely lose the signal. Professional audio designers either work with the director to stage movement to prevent this interruption, or install duplicate receivers in two or more locations. Also some inexpensive wireless microphone receivers will pick up other radio signals from nearby frequencies. Nevertheless, a wireless microphone, if well chosen, can provide good sound and allow the teacher to move around freely.

There are two considerations for microphones which will be mounted on desk or table tops. First, they must be insulated or isolated from the surface so that you will not hear paper shuffling, fingers tapping, and other unwanted sounds. There are several mounting devices available to solve this problem. Second, assuming you are using wired microphones, you need someplace to run the cables where people will not pull them loose, or trip over them, or where janitors will not suck them up in the vacuum cleaner. A conduit of some sort is the best solution. Since the electric signal traveling down this cable is very low voltage, there are normally no building code restrictions involved.

MONITOR LOUDSPEAKERS

Some would argue that the loudspeaker is the single

most important component of the audio system. Without good speakers, you will never realize all the capabilities of the system, nor the creativity of the sound designer who produced the signal. As microphones hear better than most humans, good loudspeaker systems reproduce sounds better than many of us can perceive. Due to the shape of the human ear and ear canal, we perceive the mid-range of frequencies (pitches), from about 500-4000 Hertz, as being louder than frequencies either above or below that range. Since humans are capable of hearing a range all the way from 20-16,000 Hertz, speaker systems are usually designed to accentuate the sounds we do not hear well more than the ones we do. The most basic speaker system, called two-way, includes a woofer for low frequencies, a

Murray State University, Bob Lochte

Loudspeaker mounted where the directional microphones cannot hear it.

tweeter for highs, and an electronic crossover network that channels frequencies above or below a set point to one or the other of the speakers. You can also use the tone controls on your receiver to emphasize bass, treble, or both.

Most human speech sounds are in the mid-range, with vowel sounds generally toward the lower end and consonant sounds toward the upper. Thus, even low fidelity audio channels, like voice-grade telephone lines, will do a decent job reproducing the voices in your classroom. They will, however, sound tinny over the loudspeakers. Even though most communication in the class will be voice, you may choose to use recorded music, or videotapes with professionally mixed soundtracks, as instructional resources. The speaker system should be capable of reproducing these sounds faithfully.

Students at the distant classroom must be able to hear and understand you. Depending on the size and configuration of the classroom, you will probably need two or more loudspeakers for the students and one at the teacher's console so that you can hear them well. The closer the speaker to your ear, the better you will hear. In the best of all possible worlds, each person in the classroom would have an individual monitor or set of earphones. We shall discuss these options more in the section on classroom design.

Having microphones and speakers in the same room presents some problems, but none that are monumental. If the microphones hear the output of the loudspeakers, you will create a feedback loop that will at the least interfere with audio transmission, and at worst send echoes reverberating down the line. You can avoid this problem by mounting the speakers in places away from microphone pickup patterns, an easier task if you are using directional microphones, and by adjusting the output volume on the speakers and the input level on the microphones. Fortunately, we all have marvelous sound monitoring devices, called ears, that allow us to perform such technically intimidating tasks.

Murray State University, Rod Freed

Microphone mixers and a monitor speaker in the ITV classroom.

OTHER AUDIO AND CONTROL EQUIPMENT

Most interactive television classrooms will have an audio mixer, control console, or board. This is a very useful and versatile piece of equipment. It allows you to monitor individual audio sources, adjust their individual volumes, mix two or more for a composite sound, and route the chosen output to the distant classroom. The really nice feature of mixers is that all of the operations can be automatic so you don't have to fool with them!

For example, in a classroom with 12 microphones, the console might turn each one on whenever someone starts speaking and adjust the level so that you can hear the voice at the distant classroom. These two features are called gating and automatic gain control and are desirable options.

Automatic gain control can also prevent distortion, caused by sending too much signal down the line, but cannot filter out noise, which is natural but unwanted sound. An alternative to gated, or voice-activated microphones, is individual on/off switches on each one, but the students must remember to turn them on and off.

Automatic gates and gain controls respond rapidly but not instantly. There is always some perceivable delay, usually losing the first one or two syllables. If you are using compressed video and audio, there will also be a delay in digitizing the signal, transmitting it, and expanding it on the other end, sometimes as much as two seconds. Therefore you must establish a protocol of waiting briefly to respond so that you can have a clear channel. Also, everyone in all classrooms must stay quiet unless they have a question or comment, or you call on them to speak. This protocol will require explanation and attention on your part because it is natural for people to talk simultaneously in conversation and unnatural for students to keep quiet.

The audio console will also have inputs for other sources. An audio cassette player is necessary if you will be using narrated filmstrips and slide shows. The audio line from a VCR can come through another input. If you are using music, you may want a CD player or turntable. All inputs on the console are not the same electrically. Even though the plugs fit, the signals may not match the circuitry. Always plug microphones into inputs labeled "Mic," turntables into inputs labeled "Phono," and everything else into inputs labeled "Line." If in doubt, ask your technician for help.

CONCLUSION

When you think about it, you probably communicate with your students more by sound than by all other means. You need to pay attention to the audio system in the interactive television classroom in the design phase, devote time to mastering its use, and suggest necessary modifications and improvements, many of which will only involve moving pieces of equipment. If you cannot hear and understand your students and they you, it does not matter what they can see.

CHAPTER 4

PREPARING AN INTERACTIVE TELEVISION CLASS

In television production, the preproduction phase makes or breaks the quality of the final product. In this phase, the producer makes decisions on how to allocate human, technical, and financial resources, approves a scenario that becomes a script after several revisions, and gathers all cast and crew for rehearsal. These activities can go on for months. Once the team is working well together, then the tape rolls, and the actual production begins. Later, the postproduction specialists edit the video, create special visual effects, mix, dub, and sweeten the audio track, and put the program into final release form. Regardless of the level of technological sophistication, if the human beings involved do not use their acquired skills, plan every step in detail, communicate, delegate, and work together, the production IS AWFUL. Look at the credits of a television program and you will see how many more people than machines are involved.

You should approach planning an ITV class in the same manner. The more time you can spend on this effort and the more details you can provide to your team members, the more you will be satisfied with your students' progress. Although it may seem that you are working in isolation, there are other people involved in your class. You have a supervisor, sort of an executive producer, who expects certain results. You have office personnel and teacher's aides who are responsible for distributing any printed material you provide. There is a librarian who can help with reference material, and a technician who may also work wonders if you describe in detail what audio or visual resources you want.

And don't forget your students who are your best human resources, and most incisive critics. To your basic skills of subject mastery and teaching ability you must add time and personnel management.

ORGANIZING YOUR THOUGHTS

In order to manage the efforts of others, you must have objectives. For a first ITV class, these might be to foster student learning experiences consistent with traditional classes and to master teaching in the ITV classroom. You may add to this list as you see fit so long as you plan each ITV class you teach with objectives firmly in mind.

The next step is to decide how you will determine whether or not you have achieved the objectives. In the case of student learning, comparing grades will suffice. How to evaluate your own ITV teaching ability and comfort with the medium can involve independent review, student evaluation, personal reflection, or a combination. Whatever you choose to do, do it and use the results productively.

Now that you have objectives, go back and look over your previous lesson plans or syllabuses. Think about offering this subject as a correspondence course. Logistically, ITV resembles this type of delivery more than it does your regular classroom. You will need to restructure your syllabus along a timeline based on your deadline dates, the time it takes students to learn certain concepts or to complete assignments, the time it will take you to communicate with your team members, and the time it will take them to do what you need done. This timeline should take two parallel tracks, one for students and subject matter and one for classroom management details. You will find that the latter starts before the former. Leave plenty of room because you will have many details.

CLASSROOM MANAGEMENT	CLASS SCHEDULE
Week of July 12 1. Complete syllabus and deliver to site manager. 2. Schedule conference with artist for graphics design.	
Week of July 19 1. Prepare handouts, tests, and out-of-class assignments for the first four weeks of class. 2. Deliver these items to the site manager.	
Week of August 3 1. Schedule conference with artist to proof graphics. 2. Schedule conference with site manager and teacher's aide to discuss the syllabus.	
Week of August 17 1. Collect graphics and other visuals. 2. Check syllabus, handouts, and other duplicated material. 3. Schedule conference with site manager for last-minute details.	
Week of August 24 1. Prepare handouts, tests, and out-of-class assignments for the second four weeks of class. 2. Deliver these items to the site manager.	1. Meet first class, pass out syllabus, and discuss the schedule. 2. Teach the students how to use the ITV equipment. 3. Start teaching.

Sample timeline for class organization.

Now start working backwards from your deadline dates for exams, papers, etc. Certainly you must consider subject matter to be covered in advance of these dates, listed in your student timeline. But how are you going to deliver these exams to the distant sites, how are you going to have students return their work to you by the deadlines, and who will supervise this process? These details need to go on your class management timeline. You may choose to FAX papers to the distant sites. Remember it takes time to transmit the document and duplicate it for each student, time that can be spent in other classroom activities. One instructor I know brought in a 29 page handout to be FAXed to two distant sites and copied for 14 students. The process took almost an hour during which time he stalled because he had planned his class around that handout.

Although FAX machines and computers linked via modem are more immediate, low tech distribution, like the US Mail and courier services, works reliably if you plan enough time. My rule of thumb is to plan on mail distribution once a week with a minimum of three weeks' lead time to print and copy the items to be mailed. Mail is an excellent and inexpensive way to get assignments, papers, and exams back and forth. You need only specify a postmark date as the deadline and mail back the graded papers. Another promising alternative is using computer bulletin boards and shared software to transmit these materials electronically. Again you can specify a log-on date and time deadline, but the student can get feedback sooner in his or her own computer mailbox.

If you were teaching a correspondence course, you would package the material conveniently for distribution. Before the semester begins, you would have all tests, assignments, handouts, and a detailed schedule for the students printed and ready for distribution on certain dates. You would label each item for subject matter, but you would also catalog each by number or letter so that the person handling distribution could mail "Handout 17" on

March 4 or "Quiz 2" on September 22 with less probability of error. You should follow the same procedure with an ITV class, especially if you are the person handling distribution.

ORGANIZING INDIVIDUAL CLASSES

Now you can begin to think about the other track on your timeline, the important one that deals with the students. Based on your class management decisions, you may need to adjust the lesson plan, for it may be logistically difficult to follow in some instances. For example, you may have had a paper due or an exam scheduled on a Monday when the post office is closed. There may also be dates when personnel are not available, or the schools that share the ITV network may have different schedules. You must find out all of this in advance and adjust the lesson plan accordingly. Another consideration is the availability of textbooks, reference materials, classroom supplies, and teacher's aides at the distant sites. It is worth your time and the price of a few phone calls to communicate with office and library personnel at the various sites and build a personal network so you know whom to contact for help. At the same time, you can check on the location of FAX and copy machines, and computer terminals, and find out who will be around to operate these devices. Based on the revised lesson plan, you may need to make some changes to the class management plan.

If you have made a detailed plan, have communicated with all people involved in its success, and monitored their efforts, you have come a long way toward achieving your objectives. The students will be impressed with your level of organization and will respect you for it. Plus you will feel more comfortable in the classroom realizing that the details of daily management are taken care of and that you can concentrate on teaching and learning. You will notice that up to this point I have talked very little about ITV, but all good things must come to an end.

You must begin to think visually for an ITV class to work effectively. To do this, go back over each daily lesson and jot down some ideas for how to depict or diagram the major concepts. Photographs, charts, maps, filmstrips, video, and film are possibilities, but so are simple outlines and concise tables of data. Theoretically, any visual or graphic material can be transmitted on television. The next step is to locate these visuals if you can or create them if you cannot. Whenever you encounter a visual aid you would like to use, look at it on TV. It may need to be altered or even recomposed to read well on the screen. Transparencies for overhead projectors or tables on a printed page, for example, are usually composed vertically and are not in proper aspect ratio. The type is also too small. In fact, I find many transparencies illegible even when projected. For ITV, you will at least need to have glossy cells copied onto plain paper to avoid reflections into the camera.

When you look at these visual aids, make sure you view them on the poorest monitor at the greatest distance to approximate the worst experience a student might have. For tables and text, you will find that the camera can transmit about nine single-spaced lines with about four words per line. The text will be most legible if you use all capital letters. If you need more information or words per page, plan to distribute a handout and outline the key points on one or more visuals. Another common problem is the chart or graph whose diagrams are legible but whose text is too small. You may also want to duplicate this as a handout so that students can read from the paper what they cannot see on the screen. Some photographs and slides will be out of aspect ratio. For these you can use camera zoom and movement to crop or isolate areas that tell the story and look good on television. Good visuals will illustrate your lessons, focus the students' attention where

This text is prepared to fit within the frame of a television picture. To see how well it corresponds to the aspect ratio, place it in front of a camera and see if you can read it!

For preparation of text graphics, stick to nine single-spaced lines with 4 words, or 22 characters, per line.

you want it, and break up the monotony of staring at a television screen for an hour or more.

The other major adjustment to your lesson plan for ITV will be to design interactive exercises. The first step will be to teach the students to use the ITV equipment, and this should be part of your activity the first week of class. You will find that they become more comfortable more quickly in the environment and that they will feel like a partner in the learning process. Plus it will be an enjoyable learning experience for most of them.

The simplest interactive exercises will be verbal. Roll call can be an icebreaker if you wish to get everyone involved at the outset. Plan this as part of your daily lesson, then continue to converse rather than lecture for the rest of the class. Each student could make at least one presentation from the teacher's station. You can also design projects with students from one site composing questions that students at another would answer. Make sure that they

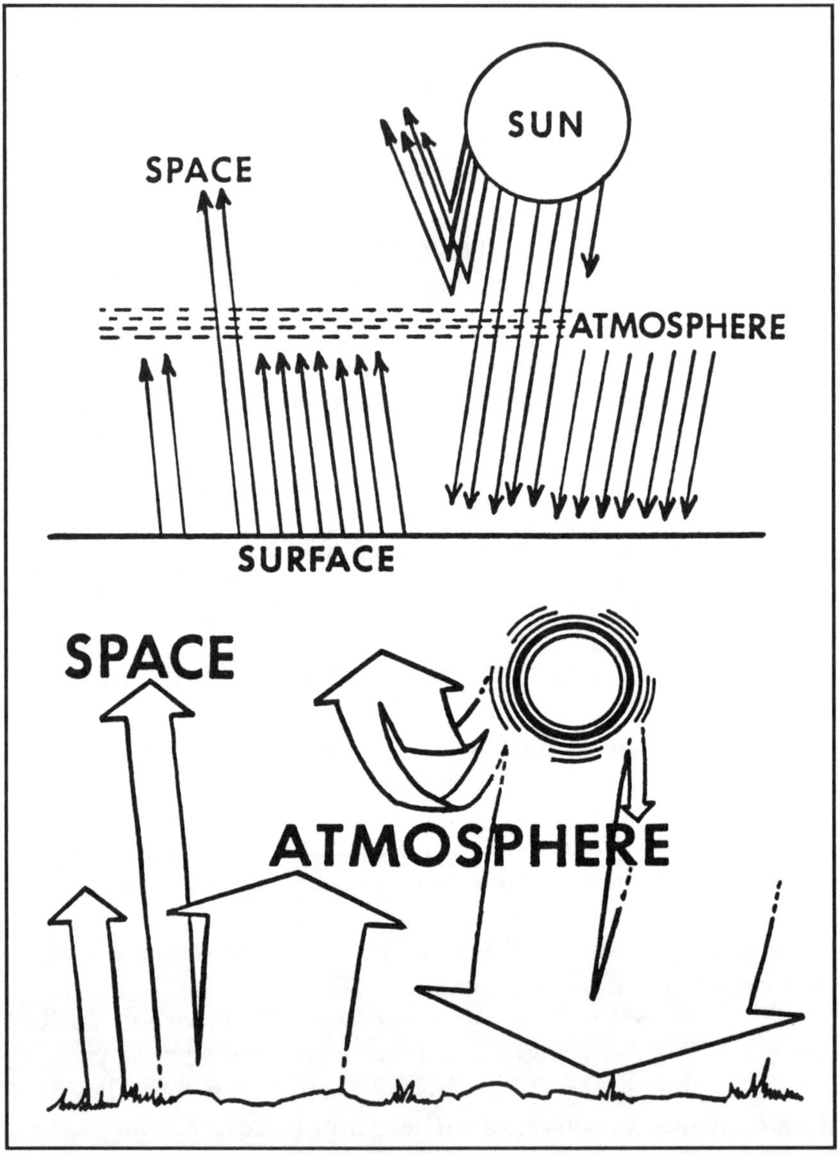

Even though both diagrams are in proper aspect ratio, the one below reads better on television because the labels and the graphics are bold, not busy.

ask students by name to respond to help foster personal relationships and camaraderie. Remember they are part of your team that will make this ITV production a success. Planning exercises that force them to talk will help stimulate the interaction that is necessary to achieve your objectives.

Let me quickly interject a point. Do not expect to see non-verbal cues such as facial expressions on a regular basis. You simply will not have a good enough view of individual students. On the other hand, if this were truly a correspondence course, you would not see them at all.

You can, however, plan interactive video exercises. Some can be elaborate, using computers with modems and shared software, or PC-controlled laser discs for a video game environment. Lacking time, budget, resources, or expertise, you may want to resort to exercises designed for the chalkboard or the overhead projector. All you need to do is to send copies of the visual layout to each site and have a student place it under the overhead camera. This can be a fill-in-the-blanks, or a complete-the-diagram, or a problem-solving exercise laid out in steps. Then you can alternate between students and sites as each adds a piece to the puzzle. Make sure you send enough copies of the format so that you can start fresh if the picture gets messy. The book by Cyrs and Smith, in the bibliography, has many helpful suggestions for interactive exercises.

On your student timeline, you need to allocate time for conferences just as you would have planning periods or office hours for these purposes. If the ITV rooms, your time, and the students are available after class, this is probably the best choice. You will have plenty of details to attend to before class just to get started on time, so this is a poor option.

Conferences are desirable to build personal rapport with the students. Remember that yours are not the only students and classes and that conference time must be scheduled at both sites to assure you will have both rooms.

Also, make sure that all other sites are off the network so that your conference is confidential. The most practical way to handle student-teacher conference time is to put it on the master network schedule so that the facilities will be allocated for that purpose, then schedule individual conferences during these blocks of time. Put these times on your timeline and communicate this information to the students. When there are no conferences scheduled, technicians can use this time for maintainance, installation, testing, or demonstration and training. Although video conferences are preferable, a telephone call can be just as effective and easier to arrange.

CONTINGENCY PLANS

The best contingency plan is to become so familiar with all the ITV equipment that operations become second nature in the classroom. This will require practice, practice, and more practice. You also need to rehearse some lessons using the clock because your timing will be different from that in a traditional class. Some lessons that are well visualized will move along very rapidly, for visual communication can be both complex and quick. Others that entail using many different pieces of equipment may move slowly if you fail to plan enough setup time. Just as your students will be impressed with your organization, they will be truly amazed if you are a whiz with the equipment, or at least do not get bent out of shape when something goes wrong.

When things go wrong, you need a backup plan, not only to restore communication with the distant sites, but also to restore continuity in the lesson plan for that class. The first element of this plan is a set of three documents: a list of the phone numbers of key persons and all classrooms; a list of items to check to see if transmission can be restored easily; and more detailed operating instructions for individual pieces of equipment. Preparing these lists may not be your responsibility; however, it is in your best interest to see that this task is accomplished and that the lists are distributed to all classrooms. Check for these the first day of class and let your students know what the procedures are when you lose video, or audio, or the signal altogether.

It will also be helpful if you practice using whatever backup transmission system you have in place. Early in the semester, plan a lesson where you cap the cameras and just use the audio signals. Or shut down the system and use the speakerphones, then restore transmission with the help of your students. This can even be an exercise exploring the importance of visual communication, or group problem solving, or coping with the frustration caused by gadgets that are supposed to make life more fun.

CONCLUSION

Because ITV has been sold as the next best thing to being there, we tend to compare it to the traditional classroom. In fact, ITV classes bear little resemblance to the experience of having a live teacher in the room with live students. But ITV is equally unlike one-way televised instruction. We are just now learning what this environment is and how to use ITV to promote education. As with any class you teach, or any complicated endeavor you undertake, planning each step in detail, and rehearsing as many as possible, will hasten your sense of competence with ITV. By planning to teach your students about the technology as part of the subject matter, you will enrich the learning experience and make them your partners in achieving the objectives you have established for them.

CHAPTER 5

TEACHING AN INTERACTIVE TELEVISION CLASS

From the previous chapter, you may surmise that the optimal situation for an interactive television class is a specialized subject with small groups of students at each site who are stimulated by the material. This seems to be a correct assumption. ITV is actually more cost effective in this type of class where no one site would have enough students to justify the expense of hiring an instructor. Frequently you will use ITV in such an environment.

Often, however, you will deliver a class with full rooms at each site and required subject matter which many students are taking only because they need these credits to graduate and this class is easier than an alternative. This situation strains the effectiveness of the ITV technology. It becomes difficult to have enough cameras to see students well or enough microphones to hear them. Likewise, there never seem to be enough (or large enough) monitors for the distant students to see you. ITV may provide additional technical reasons for the poorly motivated student to become even less so. This situation will be frustrating to you because you will tend to blame yourself for environmental problems which may be beyond your control.

Nevertheless, there are ways of dealing with all problems that ITV presents short of canceling a class. Most problems we encounter in life have no solutions, only ways to live with the situation with less frustration. ITV is no different. This chapter will provide you with some things you can do to make the ITV class run more smoothly, some things your students can do, and some anecdotal comments, both positive and negative, from ITV instructors and their students about their experience with the medium.

CHECKLIST FOR ITV INSTRUCTORS

1. Dress properly for television. Remember how the camera sees the world. Avoid high contrast in your wardrobe choices. In general, avoid white shirts and blouses with black or navy coats, dresses, and sweaters. Stick with neutral shades that are color-coordinated. Also consider the contrast between clothing and your skin tone.

Avoid closely figured patterns such as small checks, herringbone, glen plaid, seersucker, or close-striped prints. Even on an accessory like a necktie or scarf, these will create annoying *moiré* patterns in the picture. Moreover, shiny jewelry may create little points of light too bright for the camera to see, and may even damage the tubes of older television cameras. Plus necklaces tend to clank against lavaliere microphones.

If you wear makeup, you must again avoid high contrast. Theatrical and television makeup is an art that you probably do not have time to learn, but you can acquire some basic skills. Use a flat base that is slightly orange. Any blush should be only slightly darker than the base. Do not use eye shadow. Your eye sockets will be dark enough on camera, particularly if they are deep. Instead, use mascara and narrow eyeliner to highlight the eyes. If you have pale skin and dark hair, you especially need a warm base makeup to lower this contrast ratio and avoid the Elvira look. If you have brown or black skin, you are better off wearing no makeup at all. But remember, with darker skin, avoid lighter clothes like shirts or blouses that will be next to your face.

If all of this confuses you at first, look at the way professionals dress for television. Pay attention to their makeup,

Murray State University, Rod Freed

A typical ITV class in session, in a room which has been adapted rather than designed for the medium.

which may be so well done that it doesn't look like makeup. If in doubt, try out an outfit or look on camera and see what you think. Have some fun with this!

2. Before you begin, make sure everything is working. You are depending on many pieces of equipment, most of which are reliable or durable. Nevertheless, machines fail to work sometimes. Switch through your cameras, VCR, and other video sources to check each one. Make sure your students at the distant sites see what you think you are sending. You can test the microphones and establish an environment of interaction at the same time. Check your monitors for the pictures you want your homeroom to see. If there are problems, deal with the ones with quick solutions. Revert to your contingency plans if there are no

quick fixes; otherwise, you will waste valuable class time and increase your level of frustration. Your teacher's aide should be able to handle many of these checks for you.

3. Start the VCRs to record the lesson. Again, your teacher's aide can handle this at the homeroom and can train a student in this task at each distant site. In some compressed video systems, you may choose which site to record at each one. Make sure the VCR is recording you delivering the lesson at each site. This is a reference copy for review or for students who are absent. Recycle these tapes after two weeks.

4. Speak up and tell your students to do the same. I find that the students I have trouble hearing in an ITV class are the same ones I have trouble hearing in a tradi-

CHECKLIST FOR ITV INSTRUCTORS

1. Dress properly for television.
2. Before you begin, make sure everything is working.
3. Start the VCRs to record the lesson.
4. Speak up and tell your students to do the same.
5. Wait before you respond to students at distant sites.
6. Ask questions of specific students every 10 minutes.
7. Use a seating chart at each site.
8. Pay attention to what the distant sites are seeing.
9. Write legibly and use thick markers.
10. When transmission fails, establish your backup communication link immediately; then try to get the ITV system working again.

tional class. Encourage the students to tell you whenever they have trouble hearing you or each other. This has the double purpose of letting you know about problems and stimulating open communication in the class. If your voice is soft or trails off at times, wear a lavaliere microphone to get it closer to your mouth. Students who speak softly should sit nearest the microphones on the desktops.

At times, you may want to repeat questions, comments, and answers from students so that you are sure everybody has heard them. Serving thus as a gatekeeper fosters understanding at the expense of freewheeling discussion.

5. Wait before you respond to students at the distant sites. Many ITV systems have a time delay, some as much as three seconds, in processing the audio signal. When audio comes from two sites simultaneously, even during the delay period, the two signals cancel each other and no one hears anything. A good rule of thumb for you and your students is to count to two before you start speaking. At first, this protocol will inhibit conversation, wherein people normally talk simultaneously, but do not let it discourage you from engaging the students. You will all get used to the audio quirks more quickly if you practice this protocol.

6. Ask questions of specific students every 10 minutes. This practice encourages interaction and promotes experience with the audio protocol described above. Plus it is good instructional method to involve the students in active learning. Soon you will find that they chime in on a regular basis without prodding.

7. Use a seating chart at each site. This is a critical requirement if you have only one camera covering the entire group of students. You will most likely not be able to recognize each individual student. You can begin the first class by having the students at each site introduce themselves, and in the process move the ones with soft voices closer to microphones. If there is a small number at a site, move them to the front rows near the center so you can zoom in the camera on this smaller grouping. Also, with

small classes, each student can have a microphone, at least in theory.

Once you have positioned them for optimal audio and video, then pass out the seating chart form for that classroom. They can FAX it back to you so you can start using it that first class. Ask the students to sit in the same seat each day and to raise their hand when they have questions or need to communicate with you. You can then recognize each student by name if not always by face.

8. Pay attention to what the distant sites are seeing. The students there only see what you send them so look at your output video monitor frequently. It is all too easy to cut away to a chart on the overhead camera, talk about it, and forget to cut back to yourself. Or in a lecture class with little visualization, the tendency is to leave the camera on yourself. Remember that the students' eyes will drift every ten minutes or so, and that watching a small box for extended periods of time is difficult. If you are lecturing, cut away to the shot of your homeroom students every three minutes just to break up the visual monotony. Or as suggested above, ask questions of specific students at the various sites.

If you are not sure what the distant sites are seeing, ask. If you establish an atmosphere of open communication at the outset, perhaps your students will begin to tell you when they cannot see something, or when they think the camera should be back on you. This feeling that you and the students are learning ITV together can be infectious and can help promote learning the subject matter as well.

9. Write legibly and use thick markers. A piece of paper under the overhead camera will serve as your blackboard and is easier to use. You can write in your normal writing posture and do not have to worry about standing in front of the text or turning your back to the students. The fat lines of felt tip markers show up best. Print in capital letters for best legibility. Use the horizontal format proportional to the aspect ratio of the television screen and leave a border

area to compensate for the edges lost in transmission. Again if you are not sure whether students can see or read what you are transmitting, ask them.

10. When transmission fails, establish your backup communication link immediately; then try to get the ITV system working again. The students at the distant sites have no idea what is going on in this situation. At bare minimum you have lost their attention, and the technical problem has disrupted your class. Pick up the phone and call first to let them know what has happened. A backup system consisting of a conference telephone call, speaker phones, and FAX machines at each site will suffice to continue your class if the ITV system requires major repair. Otherwise, stay on the phone until you restore transmission. Use your judgment to determine when and if to give up on ITV for that particular day's lesson.

CHECKLIST FOR ITV STUDENTS

1. Make sure you sit in your assigned seat. The seating chart that you fill out the first day will help your teacher identify and recognize you. If you need to change seats for some reason, let your teacher know so he or she can change the chart.

2. Identify yourself by name and class site when you want to speak or ask a question and wait for your teacher to recognize you. Then raise your hand so he or she can see you. This procedure assures that the teacher knows who is speaking and that your comments will be heard by everyone.

3. Pause and count to two before you speak. There is a delay in the audio signal. This pause provides enough time to clear the line so that another signal will not cancel what you have to say.

4. Speak up and talk into the microphone. You want your teacher and all the students at all sites to hear you.

5. Don't talk or make noise in class. Your voice or the sounds of papers rustling or fingers tapping on the desk will turn on the microphone and interfere with the class.

6. Clear away any books and papers you don't need from the desktop. These look messy on camera and can block your face or cover the microphone by mistake.

7. Any time you cannot hear or see anything for any reason, let your teacher know immediately. He or she may not realize that you are having these problems and that something needs to be done about them.

COMMENTS FROM INSTRUCTORS

At Murray State University, we have interviewed our ITV instructors periodically, in both group and individual settings, during the past year. While their comments are about our specific ITV program, they nevertheless reflect a wide range of concerns and issues of a general nature. I include these with some comments of my own to stimulate your thinking about the ITV environment and its impact on the learning process.

The Technology

"I wish the video and audio were a little better. I wish I could zoom in on them."

"The switching is a distraction from an interactive standpoint. As long as the instructor has to go through all that, you're going to have a presentation mentality."

"The technical arrangement has allowed me to review my own teaching."

"I often couldn't hear the students nor they me.... I can see the students at the remote site, if one can agree that a three-inch high, two-dimensional figure on a TV screen is seeing them."

"Audio seems to be the only consistent problem."

"I have learned about the camera."

"New experience. It's been fun."

Our research indicates that about two-thirds of the instructors became comfortable with ITV and desire to teach another class with the medium. Interestingly, the same proportion of students expressed a desire to take another ITV course. There is some indication that the instuctor's attitude toward the medium influences the students' attitudes, just as it does with subject matter.

The Teaching Experience

"I have devised more audiovisual aids that contribute to course versatility. This method forced me to be more organized both with course presentation and in distributing materials."

CHECKLIST FOR STUDENTS IN ITV CLASSES

1. Make sure you sit in your assigned seat.
2. Identify yourself by name and class site when you want to speak or ask a question, then wait for your teacher to recognize you.
3. Pause and count to two before you speak.
4. Speak up and talk into the microphone.
5. Don't talk or make noise in class.
6. Clear away any books or papers you don't need from the desktop.
7. Any time you cannot hear or see anything for any reason, let your teacher know immediately.

A good basis for discussion the first week of class.

"I have to proceed more slowly...hard for me to get all the material covered."

"Need more time to prepare, perhaps the semester before."

"ITV is a cool, artificial medium and is a poor substitute for the real thing....If ITV does not improve teaching, why use it to teach?"

"The biggest drawback for me with the ITV courses is the lack of student contact. My 'strokes' as a teacher come from personal contact with students. That is why I travel to remote sites as often as I do."

"I adapted teaching materials for ITV."

Learning to teach on ITV will be hard work. You simply will not be able to use your old lesson plans without some changes for the new environment. And you will have to go through the tedious and sometimes painful process of reviewing videotapes and critiquing yourself. Traveling to the remote sites is desirable, at least one visit per site per semester. If this is not possible, try to schedule a one-on-one video conference with each student at the distant sites. The technology should allow this, but again it means more time commitment from you.

The Students

"Students are not as talkative."

"At the remote site, the student-initiated critical input was less than 10 percent of that at the origination site."

"Students at the site where I am seem to contribute more to class discussions. Verbally they have indicated they prefer my presence."

"The use of an outline for the students for each lecture is extremely helpful."

"Students who wouldn't otherwise take the course at all take it because of the ITV remote site."

The dilemma here is that, for the student, ITV presents

opportunity at the cost of direct contact with the instructor. Perhaps over time the ITV class environment will become more natural for both teachers and students. Carefully choosing courses, instructors, and even students might help. It is clear, however, that the instuctor must be well organized and must encourage, even require, interaction and communication with students at the remote sites.

COMMENTS FROM STUDENTS

As part of the ongoing ITV evaluation at Murray State University, we also measure student attitudes toward the ITV classroom environment, the ITV learning experience compared to a traditional class, and their desire to take another ITV class. For the classroom, the technical variables of being able to hear and see are understandably most important, as well as feeling at ease in the environment. Instructor variables, including organization and interaction with students, combined with a comfortable classroom contribute to the feeling that the student learns as much as in a regular class. Students who adapt to the environment and who do not feel shortchanged are likely to take another ITV class. The following are some comments from these students about ITV.

What They Like Best

"The diversity of having more students."

"Review of lectures is easy with library tapes."

"Being able to take the course."

"Being on the edge of new technology."

"Having a real person to interact with as opposed to a videotape."

"I enjoy the uniqueness of the experience."

"The small class size has been ideal. The group has developed a good relationship."

"Video can be fun but it needs a game type of approach, like Jeopardy, to keep your attention."

Convenience and opportunity were the most frequent responses. Still it is interesting to note the number of students who said they enjoyed being part of something new and different. Whether this excitement makes them better learners remains a question, but I always prefer an attentive audience for whatever reason.

What They Want Changed

"Make it more like being there."

"Better control of sound and larger TVs in front of the classroom."

"More personal access to the instructor."

"A more adequate blackboard for the math classes."

"To not have the instructor so overwhelmed by the technology. His continual focus on this point interfered with the class time."

"Have demonstrations to allow students basic hands-on training so panic doesn't erupt when there is a malfunction."

"Check sites to see that students at each site have equal learning opportunities."

"Getting assignments to one place or another more quickly."

We tend to focus on the technical aspects of ITV because it is new and, to some extent, difficult. The students, however, consider the entire experience of an ITV course. These suggestions, and those in the next section, contain many basic human concerns that emerge in any classroom setting.

Teaching an Interactive Television Class

"Do not get into side discussions with the group where he is if the other site cannot hear."

"She calls my name and makes me talk."

"He is always making sure we understand what he is talking about and that we are hearing him."

"Provide more legible copies of daily class outline."

"It's very hard to get in touch with the instructor when she doesn't have an office or live here."

"Asks for questions regularly and encourages class participation. Also comes early and stays late."

"Involves each student in participation, asks each person to present in the 'teacher' spot."

"It's a good concept and I think it'll work a lot better once students get used to this form of teaching."

The students appreciate the instructors who care about them and whether they are learning. It also seems that the technology is not as important as the interface between it and the human beings involved.

CONCLUSION

Interactive Television is an experimental medium for distance learning, but it does not alter the basics of teaching. Knowledge of subject matter, preparation, organization, communication, interaction, and concern for students remain crucial to the learning process. Good teachers may have a bad experience with ITV, but bad ones will always fare poorly in the medium.

CHAPTER 6

EVALUATING AN INTERACTIVE TELEVISION CLASS

In the motion picture *The Gods Must Be Crazy*, the pilot of a small airplane casually tosses an empty Coke bottle out into the Kalahari wilderness where it lands near a bushman. Thinking the bottle a gift from the gods, he brings it back to the village. All members of the tribal community are astounded by this technological marvel that is harder and smoother than any bone or wood implements they possess. It also glistens in the sunlight and emits music when you blow on it just right.

Then something very disturbing happens. Everyone in the village finds that he or she has an important need to use this gift from the gods. They develop a sense of personal rather than community possession. Soon jealousy, bickering, and even physical violence, all hitherto unknown among the bushmen, erupt on a regular basis. The tribal elders conclude that the Coke bottle made life in the village more complicated and stressful than ever before. They decide that the man who received this gift should return it to the gods as soon as possible. Thus both modern and primitive societies reject the "Coke bottle" technology but for different reasons based on cultural context.

In many ways, educators are the gentle bushmen of modern professional society, preferring to remain isolated from communication technologies rather than complicating their lives unnecessarily. Ellen Dempsey, president of IMPACT II-The Teachers Network, writes:

> Most classroom instruction, school buildings, and school managerial systems haven't changed since the creation of public education. A school teacher from 1890 could step out of a Winslow Homer painting and feel right at home teaching the same curriculum in a 1990 classroom. What business or other enterprise could exist exactly the same way for 100 years?

Instructional television provides an excellent example of this phenomenon. Despite the construction and promotion of educational television stations, networks, program cooperatives, and various closed circuit or VCR-based distribution systems over the past four decades, only twenty percent of U.S. schools use the medium for instruction. Fewer than seventeen percent of school teachers have had any training in the use or purposes of instructional television.

In short, television has not been proven beneficial in the cultural context of the classroom. Bud Koontz suggests that colleges and universities especially have failed to adopt instructional television because administrators and faculty have not understood the process of innovation diffusion as described by Everett Rogers.

EXPANSION OF DISTANCE EDUCATION

Meanwhile, as transportation and communication technologies have evolved in the past century, these technologies have been used to bring education to people in widely dispersed geographical settings. From correspondence courses, to educational radio and television broadcasts, to audio- and videocassette series, to today's promise of direct broadcast satellites and fiber optics networks, all distribution media have been judged reasonably effective, based on two primary evaluation criteria: the convenience for the student, and the supposition that any education is better than none at all. Thus technologies that have not been adopted in the classroom have proved useful in distance education projects.

As education approaches the twenty-first century, distance learning is becoming an integral rather than a secondary consideration. In 1987, fewer than ten states promoted distance learning. Now virtually all states have distance learning projects in operation or in the planning stage. In the author's home state, the Universities of Kentucky and Western Kentucky, as well as Murray State University, are actively marketing graduate degree programs and upper-level undergraduate courses off campus. All are using some form of communication technology to get courses to the students and reduce the cost of transporting faculty to the distant classrooms. Elementary and secondary schools, especially in poor, rural counties, view distance learning as an opportunity to expand curriculum and share instructional resources.

Advances in distribution media technology, moreover, have made possible live interaction using audio, video, computer data, and graphics. Rather than the familiar one-way medium, television can now function as a two-way interpersonal communication device, more like the telephone.

The live classroom can expand to include students at several locations simultaneously in an environment where everyone involved can see, hear, and communicate with everyone else.

Interactive communication technology in effect merges the concepts of the traditional classroom and distance learning.

A NEW LOOK AT EVALUATION

Two questions arise. First, will educators who have rejected communication technology advances in general and instructional television in particular adopt two-way, live interactive media? Second, how can educators evaluate distance learning projects involving these new media to see whether they work? This chapter will concentrate on the latter. It is the author's opinion that properly designed evaluation research provides project participants the opportunity to modify the project itself and thereby improve utilization. By answering the second question, the answer to the first should become readily apparent before the end of a pilot project, especially one that involves humans encountering innovative technology.

Evaluative research designs need not be complicated to be functional. The key to success is planning the evaluation as an integral part of the project, analogous to a feedback circuit in electrical engineering. A complex electrical device, such as a VCR, has both electronic and mechanical components. Servos are control mechanisms that keep all circuits, motors, armatures, etc., running synchronously. The feedback circuits provide the servos with the information to make the necessary adjustments without having to shut the device down. In distance education, evaluation is a feedback circuit that can keep any project functioning properly, if the data are used in a timely fashion.

EVALUATING ITV AT MURRAY STATE UNIVERSITY

In implementing a two-way interactive television network at Murray State University (MSU), the author devised a five-phase evaluation procedure as follows:

1. Establish evaluation criteria.
2. Choose evaluation team.
3. Design research project.
4. Communicate.
5. Utilize findings.

The first three phases were accomplished as part of the initial plans for network implementation. The last two are ongoing efforts in the pilot project itself.

Alan Chute established purposive criteria for evaluating distance education based on three interrelated concepts: usefulness to the organization, user acceptance, and user comfort with the technology. The goals established by MSU were to make the learning experience as effective for the students at the distant site as in the traditional classroom and to do so with a minimal amount of behavior modification by both faculty and students. If both were achieved, two-way interactive television (ITV) would become useful as instructional technology and as a cost-effective way to deliver courses to widely dispersed classrooms.

Ultimately university administrators will determine how useful this technology is. The instructors and students, however, influence this decision by accepting or rejecting ITV during the pilot phase. These are the adopters in Rogers' diffusion-of-innovations model. At the same time, they are the decision-makers and information users who must collaborate on the evaluation process, as suggested by Patton. One evaluation team might include faculty members and facilitators, those persons responsible for training instructors to use the ITV equipment and for building ITV classrooms. A second team might bring students into the mix. Both the pilot and the evaluation will go more smoothly if there is an effort to recruit teachers who are generally positive about the new technology but who still retain a healthy skepticism. Evaluators need to constantly ask "Is this working?" and be prepared to accept "Yes" as an answer, but go on to query "How can we make it work better?" Likewise, the process is easier if all learning objectives are clear at the outset.

In the context of an individual classroom, or a small group of classrooms, there is no need for a large-scale research design involving sophisticated sampling techniques.

Murray State University, Rod Freed

The ITV classroom from the instructor's point of view.

The entire population can easily be studied and the results need not be generalized beyond the context of the pilot project itself. Thus surveys, field studies, and quasi-experiments give way to focus groups, extended personal interviews, before-after questionnaires, content analysis, and laboratory experiments. What matters most is that each member of the evaluation team determines what information he or she needs to decide whether or not the technology will foster an acceptable learning experience in a comfortable classroom environment. Then the team must determine how best to get and utilize this information. Patton has many helpful suggestions on this process. Remember to survey students before the end of the term, perhaps after midterm exams, so that you can utilize the results immediately with those students.

In evaluating the MSU ITV pilot, various research methodologies have been adopted. In a two-week workshop for secondary and vocational school teachers, the session leaders administered before/after questionnaires with both objective and subjective questions. The findings covered subjects as diverse as instructional method, technological literacy, and classroom design. Instructors have used similar questionnaire measures to gauge students' learning experiences, while others compare grades between distant and traditional classrooms. One instructor "debriefed" his class on a regular basis to encourage feedback about the ITV experience.

As Rogers points out, one problem with studying new communications technology is that frequently the innovation is not evolutionary, and there is no valid comparison to previous experience. It makes sense, therefore, for the evaluation team to communicate their ongoing experiences and findings with each other frequently. For this purpose, the MSU ITV evaluation team meets once a month for an informal conversation about the project over lunch. This setting is also an effective support group for instructors who will be teaching via ITV in subsequent semesters. This level of directed interpersonal communication is crucial to both the diffusion and evaluation processes and can be treated as a focus group for more formal data analysis.

The final phase, utilization, is where most communication technology has failed in education. By focusing the evaluation process on immediate utilization, and by being willing to adjust all elements of the pilot project to meet the evaluation criteria, failure is unwarranted. For example, during the workshop mentioned above, the facilitators learned that the teachers had virtually no knowledge of the MS/DOS-based computer that controlled the ITV classroom equipment and indicated little desire to learn how to operate it. The teacher's station was redesigned to include a simple video switcher so all instuctors could easily assimilate the ITV technology, yet those who wanted to add computer graphics could still do so. It is more practical to implement change in *medias res* to make the pilot project work rather than delay evaluation and adjustments at the expense of both user acceptance and comfort.

DISTANCE EDUCATION IN BUSINESS AND INDUSTRY

On a parallel track, the corporate and industrial worlds are adopting distance learning technology for many of the same reasons as higher education. There are perhaps many lessons in both evaluation and utilization which may be shared. But there are some important distinctions in both organizational context and culture. The overriding purpose of business is to show a profit. Any communication device or technology which facilitates that purpose will be useful to a business and presumably assimilated into its organizational culture. For the individual employee, the primary motivation for continuing education is a better job and a larger paycheck. If the process involves learning to operate a computer, or attending a teleconference, or changing behavior to accommodate technology, the financial rewards for doing so justify the effort.

Rarely do such rewards exist for either educators or students. Their motivation must come from the simple desire to learn and to do a better job. Also, the learning objectives are different. Do not assume that distance learning technology that works in the business world will work as well or at all in the classroom environment, just because some salesman says it will provide everything you need for a better life. Try it; evaluate it; utilize the evaluation to modify it to suit your needs; then make up your mind based on your experience.

CONCLUSION

The suggested evaluation procedure presented here should be germane to any pilot project in distance education where technological innovation is a factor, and is especially appropriate for ITV. It will require explanation to all parties concerned and the cooperation of administrators who usually look at evaluation as a summative mechanism provided at the end of a project. By establishing the formative feedback mechanism at the outset, however, both information users and decision makers will have access to the salient facts as the project progresses. What remains is to continue to communicate these facts and utilize the results of the evaluation in adapting the technology to meet the learning objectives' needs of both teachers and students. If the experiment will not work under these circumstances, it is probably unwise to invest further in the technology.

CHAPTER 7

THE INSTRUCTOR TRAINING WORKSHOP

Once you have made the commitment to ITV, you will be adding new classes each year. Plus you will find other uses for the ITV system in administration, in-service training, community service, and industrial development. Whenever traveling to a meeting wastes time and money, or is inconvenient, ITV can be a cost-effective alternative. Because your ITV needs and uses will change over time, and because most classrooms will feature instructor-controlled equipment, there will always be a need to orient and train new people. How to do that is the subject of this chapter.

At the outset, you will probably need to bring in a consultant for staff training. The equipment supplier can furnish someone to show you how to operate the various components, and this should be part of the purchase agreement. Knowing which buttons to push and knowing how to teach with ITV, however, are two very different skills. If you are hiring someone to conduct a workshop, make sure that person has experience teaching with ITV, and hopefully experience conducting workshops like the one you want. After this first training session, you can use the format below to conduct your own workshops and save money in the process.

In a former life, I owned and operated a radio station in a small town. Most of my employees were high school kids and turnover was relatively high. In five years, I trained about 60 disc jockeys and, in the process, developed a procedure to minimize the training period while still achieving the stated learning objective. In the first phase, I gave a thorough explanation and demonstration of all equipment, followed by two hours when the student observed operations. The second phase, which began on the second day, entailed the student operating the equipment under my direct supervision. This phase lasted two or three days. At that point, the student should have been both knowledgeable and comfortable enough to go to work in a non-critical time period. Typically, the third phase, independent operation with observation, feedback, and critique, would begin on day four and last several weeks. During this period, the person was actually employed using the acquired skills while they were still fresh.

For the workshops described below, I employ the same crash-course approach to learning ITV operations. Until you actually use the technology, you have no idea what it entails or how quickly you will become comfortable in this environment. The key objective must be to get people teaching as soon as possible. Moreover, I think it best to schedule the workshop as close as feasible to the beginning of the term so that all skills are freshest.

THE FIVE-DAY WORKSHOP

This workshop plan is designed to improve ability, not to change attitudes. It is helpful to recruit teachers who have an interest in ITV instruction, rather than those who feel coerced into a program. The latter group will probably never feel comfortable in the ITV environment. Teacher's aides may participate as well. They only need to attend days 1 and 3, but may want to audit the rest of the workshop. This design can accommodate 16 students at multiple sites.

Day 1 Activities

Morning

1. Explain how television works, what the limitations of the medium are, and how ITV is different from our normal concept of television. Show a video that illustrates this difference. (1 hour)

2. Distribute syllabus, schedule, and other printed class materials. Discuss these in detail, especially the requirements for the 5-minute and the 15-minute lessons. (1/2 hour)

3. Have students introduce themselves by coming up to the teacher's station. This allows each one to get the feel of the environment and to experience communicating with all sites at once. (1/2 hour)

4. Demonstrate all ITV equipment and explain why the classroom is set up the way it is. (1/2 hour)

5. Have each student come to the teacher's station and practice operating each component under supervision. No one will get all this down immediately, but each student should be aware of how to control the signal going out to the distant sites. (1 hour)

6. Pretest knowledge of ITV if you are doing a before-after evaluation of the workshop. (1/2 hour)

Afternoon

1. Answer any questions. If there are none, engage the students in conversation about their teaching and learning experiences in the traditional classroom. (1/2 hour)

2. Review instructional methods and learning theory. Be sure to ask for examples from the students to keep them involved and to acclimate them to the ITV environment. Topics for this review should include:

Organizing Lesson Plans
Communication and Classroom Interaction Time Management
Evaluation and Rewards

Teacher Behavior

Relate each topic to the ITV classroom. (2 hours)

3. Explain the 5-minute lesson assignment and the critique process. For this assignment, each student will make an ITV presentation, no more than 5 minutes long, on a subject of interest to him or her. It must include interaction with the students and be illustrated with at least 3 visuals for the overhead camera. Each student must provide a videotape for a review copy. (1/2 hour)

Day 2 Activities

Morning and Afternoon

1. Students spend the entire day delivering, discussing, and critiquing the 5-minute lessons. Each student will also get a videotape of the presentation and a formal critique from the workshop leader(s).

2. Wrap up the day by debriefing the students. Ask them what they have learned and how they feel in the ITV classroom now that they have been both teacher and student in that environment. (1/2 to 1 hour, depending on time remaining)

Day 3 Activities

Morning

1. Instruction on how to produce good quality graphic materials for television will include examples of both effective and poor visual aids. This session can include computer graphics if the system has this capability, but you will also need to cover this topic in a separate workshop. (2 hours)

2. There are also low-tech and logistical problems with ITV, such as proctoring exams at the distant sites, transmitting homework, assignments, and handouts, assuring access to videotapes of lessons for review, and more. Discuss these and suggest ways to deal with them. (1 hour)

3. Hold a question and answer period about the ITV technology to ascertain the students' progress. Explain the formal lesson plan for the 15-minute (or 30-minute, depending on the number of students) presentation. This should be a mini-lecture or other instructional method taken directly from an actual syllabus. The objective is to make this lesson an effective learning experience in the ITV classroom, using any and all human and technical resources available.

The presentations will be videotaped and critiqued in writing as were the 5-minute lessons. (1 hour)

Afternoon

Devote the entire period to preparation for the 15-minute formal lessons. Give each student as much individual attention and access to the ITV equipment as necessary and possible. Encourage them to experiment with the medium. (3 hours)

Day 4 Activities

Morning

Continue preparation for the 15-minute lessons, as above. (3 hours)

Afternoon

Deliver 15-minute lessons. Make a videotape and written critique of each. Discuss each one with the other students to find out what they learned and to generate feedback about ITV, what works, and what needs improvement. (4 hours)

Day 5 Activities

Morning

Continue 15-minute lessons, critiques, and discussion as above. (4-6 hours)

Afternoon

1. Complete 15-minute lessons as needed.

2. Debrief students as on day 2. Competency and comfort with ITV should be the focal subjects of this discussion.

3. Complete post-test questionnaire, which should be the same one you used for the pretest on day 1. (1-2 hours)

THE TWO-HOUR WORKSHOP

While not as extensive as the 5-day version, this workshop can introduce teachers to the equipment and the ITV classroom in a short period of time. It is also appropriate for people other than teachers who might be using the ITV classroom, and teacher's aides. The maximum number is four persons. If all participants are at one site, you will need to have an assistant at one distant site to furnish critique and feedback. Each participant should bring a videotape and prepare a 10-minute presentation illustrated with visuals that would normally be used (i.e., transparencies, charts, slides, etc.). Teacher's aides should observe these presentations.

1. Explain the limitations of television cameras, monitors, and audio equipment. Use some of the visuals provided by the participants as examples. (15 minutes)

2. Demonstrate the ITV equipment with participants looking over your shoulder or at the teacher's station at the distant site. Solicit questions. Then have each go through basic operations at your direction. (30 minutes)

3. Have participants deliver their lessons. Videotape each one for review. Provide feedback from both sites in a concise form, perhaps using the checklist in Chapter 5. (1 1/4 hours)

4. Be available for questions and consultation after the workshop.

In this instance, it is critical that the teachers review their videotapes and that they schedule some independent review of their ITV classes during the first two weeks of the term, particularly from the distant sites. The actual classes will serve as the laboratory exercises in the longer workshop.

CONCLUSION

This manual is designed to provide both general and specific information about ITV and to serve as a reference for the future. It would make an excellent supplement to either workshop described above.

CHAPTER 8

INSTALLING CLASSROOM EQUIPMENT

There are two options for your ITV classroom. The first is to build it from scratch, starting with a basic design and configuring all elements both structural and technical to suit the needs of the instructors and students. The second option is to adapt an existing classroom, lounge, storage area, or broom closet for this purpose. Most readers will take the latter choice. It is, however, instructive to look at option 1 first, see how we would build ITV in the best of all possible worlds, and then work backwards to the point of budgetary compromise that fits the individual situation.

Although I shall suggest some ways to economize, television is not cheap. It will be difficult to build an ITV classroom for less than $20,000 that will satisfy both instructors and students. And this figure does not include transmission equipment or transmission line costs. At the outset, you should establish sound and precise learning objectives for ITV students. If meeting these objectives is a priority, then you need to evaluate the cost of ITV, including the ongoing expenditure for personnel, training, line charges, and maintenance, against the cost of meeting the objectives through traditional or low-tech alternatives. No one has ever said that ITV is right for every situation.

Another point to consider is that this is a television studio as well as a classroom. To the students at the distant sites, all that matters is what they see and hear through the medium. The choice of equipment, and the placement of furniture, objects, and people in the traditional classroom may not be the same in the ITV classroom.

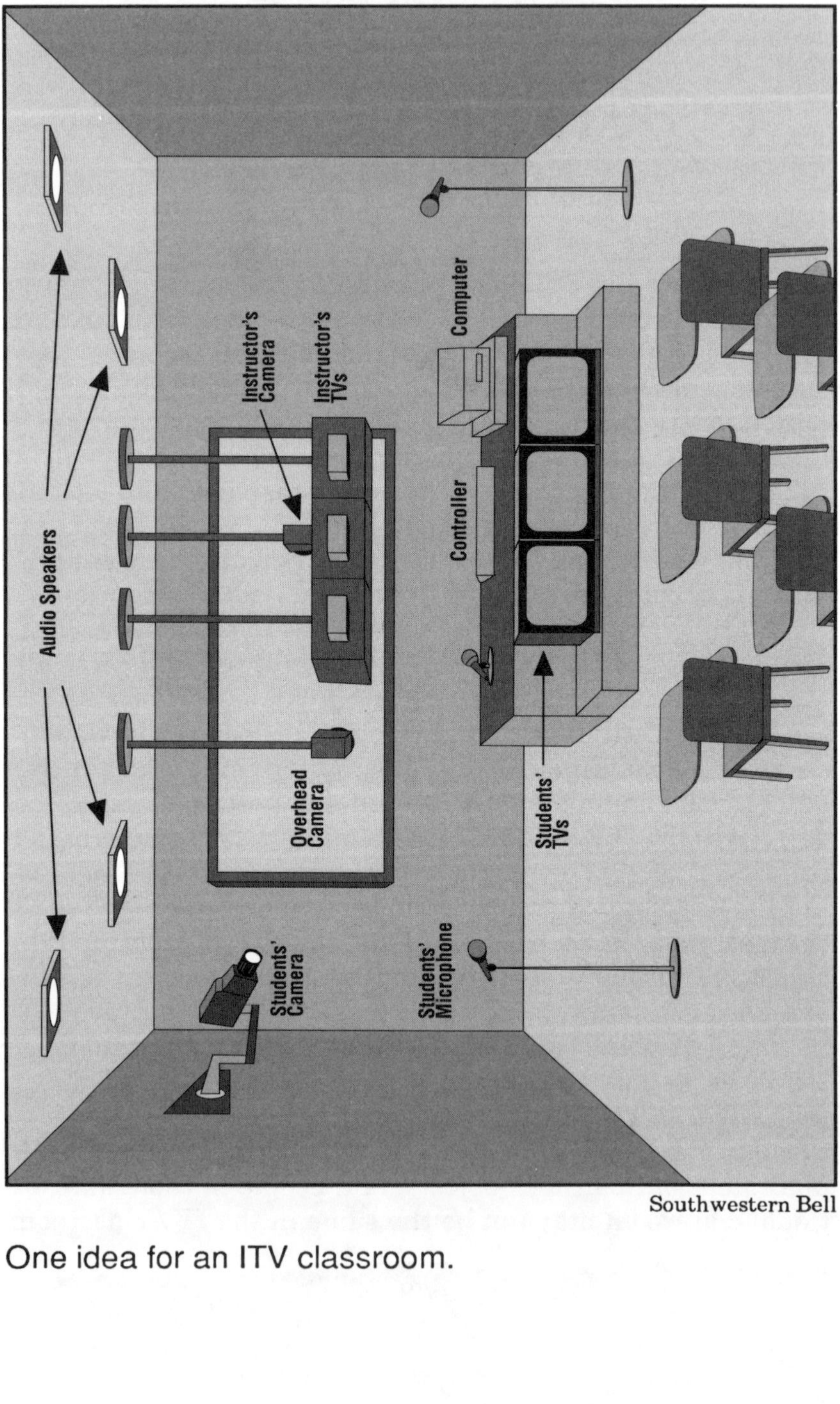

Southwestern Bell

One idea for an ITV classroom.

If you are in doubt about a choice or decision, look at and listen to the results on television. That is the only way you can tell what works. Now let's start our wish list.

THE DREAM ITV CLASSROOM

Every piece of equipment that goes into the ITV classroom needs electricity, and most of it will need cables for audio or video or both. There will be more wire than you can possibly imagine, so the first thing we need is someplace to hide it. Above the ceiling is a possibility, but if we put it under the floor we shall not require a ladder. A computer room floor with removable panels is perfect. We also need to calculate how many electrical outlets we need at what amperage and then double both figures to simplify future expansion.

Because electrical wiring generates low-frequency signals that can induce current in audio and video cables, we need to isolate all electrical wires in conduit. Local building codes probably require this as well. If you hear a constant low hum through your speakers, it is probably electrical interference from nearby unshielded wiring or an ungrounded audio cable. If you must run audio or video cables near AC current, run the cables at right angles to the electrical wires.

Because there will be so many cables and wires, most of which look alike, each one will be labeled at each end so we can trace any circuit easily in the future. Also, the technicians who install the equipment will prepare a detailed wiring diagram in a database. Having this information in computer-readable form will insure ready access and simplify future design modifications.

With the computer room floor and plenty of juice, we would want modular furniture and work stations so that we could configure the classroom differently for each class. A seminar may work better with all students and the instructor facing each other, while a lecture/demonstration class might entail a more traditional arrangement. For a class in composition or computer programming, the instructor may want to be in the center of the room with work stations for the students all around her. We should design all structural and technical components for the degree of flexibility.

There will be many pieces of equipment at the teacher's work station, and we need to design it carefully so that he or she has ready access to everything but can still maintain lines of sight and communication with all students at all locations. A wireless headset with directional microphone will assure that the teacher can hear and be heard no matter how much he or she moves around. There also needs to be a video monitor at the station that always shows what is being transmitted, and at least three more output monitors hung out in the classroom so that the teacher is always able to see what the distant students are seeing. These monitors must have at least 31" diagonal pictures and must be positioned where the students can see them. One camera would cover the teacher at the work station, and two others, equipped with motion-sensitive switchers, would follow him or her around the room.

At the work station, the teacher will need a separate monitor to show each distant site from a wide-shot perspective, with the ability to switch to a closeup shot of an individual student either by choice or by voice activated circuit. In order to maintain eye contact with the student, he or she would need a separate miniature camera attached to each monitor. Or better still, each monitor could be a teleprompter array where the image appears reflected in a two-way mirror, so the teacher looks through the picture directly at the camera lens.

Of course we would need other video equipment for resources. Certainly an overhead camera for any printed graphs, charts, or photographs, or to use for quick notes handwritten on paper. For more complex written material, such as mathematical formulas, an electronic wide screen

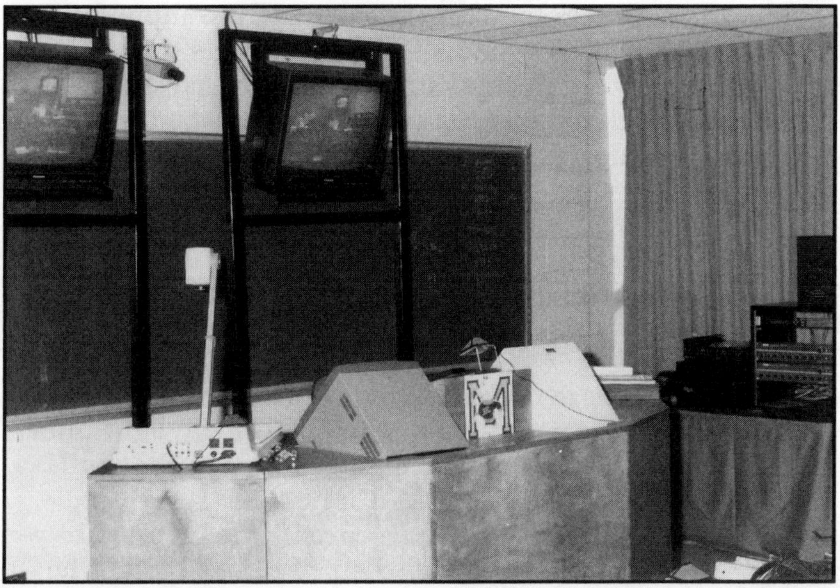

Murray State University, Rod Freed

The teacher's station in an ITV classroom, with overhead camera, monitors, and voice-activated microphones.

blackboard with a receiver at each site should suffice. There should be a separate telecine camera for slides, filmstrips, and 16mm film transmission, and at least one VCR for playback and one to record each class. Finally, the teacher must have a personal computer at the station, networked to all the sites, to his or her office, and perhaps to his or her home, equipped with the software needed as instructional resources as well as administrative support for testing, evaluation, and communication.

The student's work station would also include a headset with directional microphone, but these might be wired to keep the students in place. Certainly each work station would have its own PC and keyboard, and a monitor that could receive either classroom video or RGB computer inputs. Each station would, of course, have its own miniature camera. There would also be a camera covering the entire classroom and monitors showing both the instructor's output signal and all the distant sites, all within the field of view.

Since this is a very complex system, the teacher would need a technical director in an adjacent control room to switch cameras and originate tapes, slides, or other video sources, and a teaching assistant in each classroom on the network.

We would choose to locate this room in the interior of the building, away from any windows, since daylight interferes with the soft quartz lighting instruments that will provide an even 75 footcandles with no hard-edged shadows. Both floor and walls will be carpeted for sound absorption, and acoustic tiles will comprise the ceiling. The color scheme will be warm earth tones, with all shades reflecting 40-60% of peak white for proper contrast balance with skin tones. There will be no busy patterns and no shiny objects present, such as chrome furniture, to create unwanted highlights in the television picture. No walls will be parallel so there will be no chance for standing audio waves to reverberate and make hearing difficult. Furniture must be chosen to be ergometrically efficient and comfortable for both students and teachers for the time periods they will be present in the classroom.

The computer system would operate on a separate network and would be available as a backup link in case audio/video transmission fails. Each classroom on the network would also have a direct telephone line with speakerphone and FAX machine for a tertiary backup system.

THE MINIMALIST ITV CLASSROOM

Assuming that your school district or board is not printing money, you will be adapting existing space for the ITV classroom on a budget that is fixed and scant. The challenge is to incorporate as many as possible of the ideas above in the plan. "Where can we compromise?" is the question.

Let's start with cameras and television monitors. You will need at least three color cameras in each classroom capable of getting an acceptable picture in low light conditions. The ones designed for surveillance work will do but will not give a broadcast quality picture. If it looks fine to

Murray State University, Rod Freed

A conference, or seminar, ITV classroom with multiple cameras, wired into the voice-activated microphone circuitry, and big-screen projectors.

you, you have cut your camera cost at least in half. One camera will have a fixed shot of the teacher, one a fixed wide shot of the students, and the third will be mounted on a copy stand so that it can get an overhead shot of text and graphics. Only this third camera will absolutely need a zoom lens, but this is a desirable option on all three. All zooms are manually controlled.

There must be at least four monitors with a minimum 25" diagonal picture. Two will be placed above the teacher for the class to see, and two will be suspended over the students or mounted at the back of the room for the teacher to see. One of each pair will show whatever the teacher origi-

Murray State University, Barry Johnson

ITV instructor in class. Note the lines of sight to the overhead monitors.

nates and the other will show distant classrooms. There must be some sort of switching mechanism, either automated or teacher controlled, to choose among the various distant site feeds.

The teacher's camera should be mounted above or below the distant site monitor so that he or she has eye contact with the camera when looking at students on television. All monitors should have A/B video inputs so that the teacher can choose which will be his or her feed and which the distant site feed.

Mounting both cameras and monitors will take some trial and error. The cameras must have an unrestricted shot of their primary subject matter. The most troublesome will be the classroom shot. This camera will probably have to be in a corner to see all the students, but in this position it will not provide eye contact with them. There are two choices here. Place a monitor under this camera or arrange the classroom on the diagonal. The first option will encourage the students to watch television rather than the teacher, and the second may feel awkward.

Everyone needs an unrestricted view of the monitors and needs to be as close as possible to them. This is a major area of compromise that can only be improved by installing more or bigger monitors. The largest monitors available are projectors whose pictures are not as bright or as sharply focused as picture tubes. For the cost of one big screen projector, you can purchase three or four 25" diagonal monitors. At any rate, all television monitors have problems with ambient light. You may have to tilt them down or shade them to minimize glare from the room lights.

The teacher will need a video switcher to choose between cameras and other video sources. This can be a simple affair, but it makes sense to buy one with a few more inputs than you need now in case you want to add another camera later. If you are using compressed video, your CODEC probably has a switcher built in. Many desktop overhead cameras have auxiliary video inputs that permit expansion. Since the teacher will be doing his or her own switching, this device should be mounted on the lectern. Likewise, the overhead camera should be nearby for easy access.

There are several simple integrated audio systems on the market, designed for videoconference applications, which include multiple directional microphones and a console with automatic gain control and voice-activated switching. This is the least expensive option that you can live with. You will need one fixed mike for the teacher if she does not move around and one fixed mike for every three students. These must be placed on the desktops as close as possible to the person speaking, and isolated from ambient sounds. If you cannot afford many microphones, buy one good directional mike for the teacher, perhaps a lavaliere mount, and place one other good directional mike on a stand in the middle of the classroom so students can step up to it if they have questions. You will find any compromise in between wholly unsatisfactory.

You can try to use the television monitors as audio playback sources. They will work in small classrooms. A better and reasonably priced alternative is a small amplifier and a pair of two-way loudspeakers. Mount the speakers high on opposite walls, in corners if possible. Make sure they do not point into the pickup pattern of any microphone.

If you use the TV monitors, make sure you have more than one sound source so everybody in the room can hear. Balancing microphone and monitor levels will be trial and error, and will be different with each room.

You will need at least two VCRs, one to play back and one to record. The least expensive VHS models that are the easiest to operate are the best choices. Be aware that if both are the same make and model, the remote controls will start and stop either machine. You may want to mask

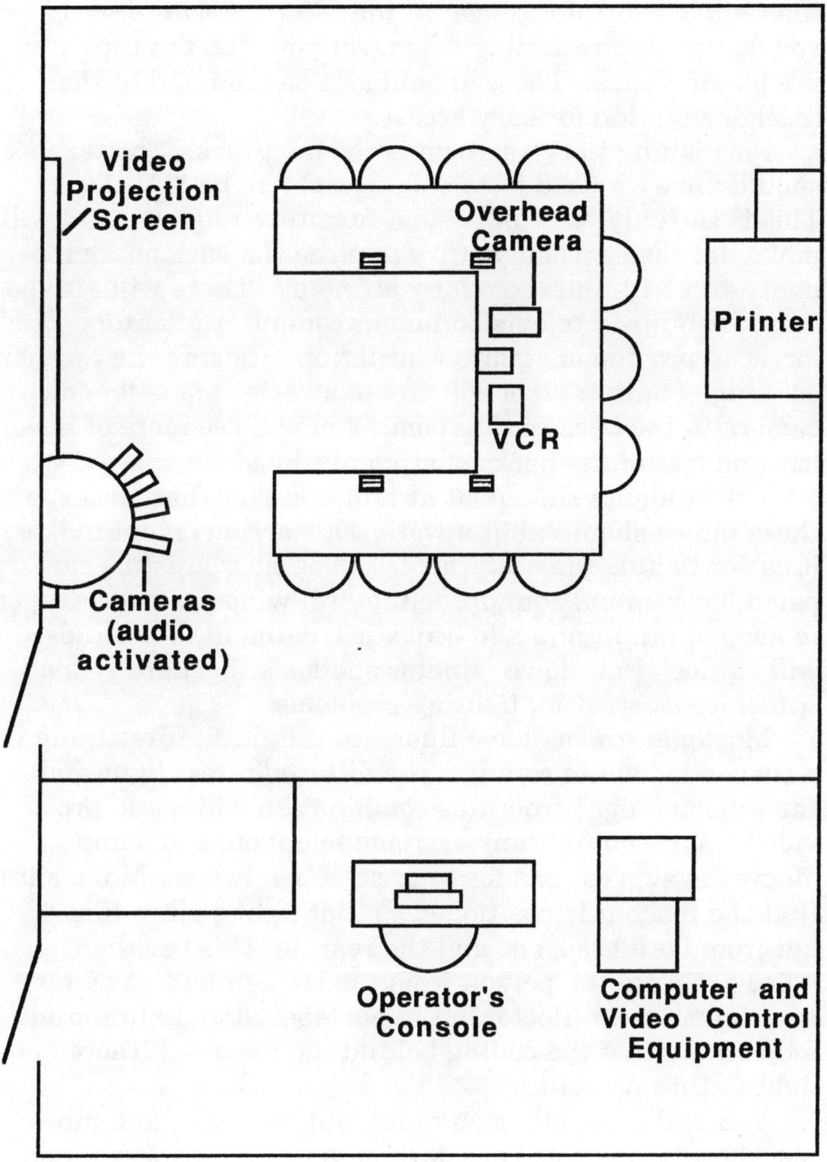

A video teleconferencing facility adapted for an ITV class.

the remote control receiver on the record machine so that you do not stop recording whenever you stop the tape you are playing back. These should also be mounted at the teacher's station for easy access.

Furnishing the classroom is the next task. The teacher should have a raised lectern designed for the ITV class. This is the only piece of custom furniture required and will make life easier when you try to place the various components and still find room for your notes. There will also be room for built-in television and/or computer monitors. Don't forget to provide adequate ventilation. Raising the teacher on a high chair or stool will create a better angle for the camera in the back of the room. You will see more of his or her and less of the backs of students' heads.

The students should sit at tables rather than desks, and these tables should abut a wall. This arrangement makes it easier to hide the audio cables. The floor must be carpeted for minimal soundproofing. All windows need drapes to keep out light and add acoustic treatment, but blinds will suffice. Pull-down window shades are generally too translucent, creating lighting problems.

Most classrooms have fluorescent light fixtures hung in a suspended white ceiling. The diffused direct light and the reflected light from this combination will work, provided you do not use any extreme telephoto shots and follow the suggestions for color selection below. Make sure that the teacher is positioned so that light strikes him or her from both the front and the rear, for this backlight helps separate the person from the background. You may want to build a reflector out of pasteboard and aluminum foil to mount on the ceiling behind the teacher if there is no light fixture present.

You will probably need to repaint the walls and mask blackboards or other very dark or very light backgrounds that will be seen on camera. Remember the camera likes low contrast settings. Use a photographic gray scale when choosing paint, fabric, and furniture, and keep all choices in the middle of the scale. Perhaps an easier way is to bring in color samples to the classroom and assemble the teachers with the darkest and the lightest skin tones. Turn the color down on one monitor so you only see shades of gray and hold up the colors next to their hands. If the gray tones are slightly lighter or darker than the range of flesh tones, then these are good choices. Check coordinating colors and fabrics the same way. When you paint, avoid small patterns that may look busy on camera and detract from the teacher. Choose desk surfaces that are light colored, so that they will reflect some light back into faces, but avoid shiny formica.

Most of the wiring will go over the ceiling. As in the dream classroom, the technicians need to label both ends of each cable and provide a detailed diagram to show how everything is connected. Any wire hanging down walls, or running along floors or under countertops, must be in conduit or otherwise hidden so that people cannot pull it loose. If theft and vandalism are problems, make plans at the outset to secure individual components as well as doors and windows.

A teacher's aide trained to operate equipment and handle light maintenance will facilitate ITV classes. Most systems will be set up with identical equipment at all sites, right down to the camera numbers on the video switcher. Purchasing equipment in volume will earn considerable discounts from suppliers. If you have one trained teacher's aide at one site, he or she should be able to solve minor problems at all sites.

A FAX machine is a useful tool for sending handouts, assignments, and tests back and forth. It is best to run this on a separate phone line and install a speakerphone so that you or your aide can remain in touch with the distant sites when the main transmission fails. Any major problems will require trained maintenance personnel. Hopefully you will have at least one such person available, perhaps in a cooperative arrangement with other schools.

CONCLUSION

As a final note, there are many choices of ITV equipment on the market, and you can spend as much money as you have.

Be aware that most of the systems that hucksters will show you were designed for business teleconferencing and not classroom use. While the components may work, the settings, people, and objectives are different enough that you will need to adapt the technology to suit your situation. Hopefully, this chapter has given you some ideas about how to do this.

APPENDIX A

USING THE OVERHEAD CAMERA

Because most ITV classrooms are equipped with an overhead television camera, usually at the teacher's desk, it is advisable to become more familiar with this very useful and versatile visual aid. Many instructors who have participated in my workshops have requested more details about this device, so I thought it appropriate to include this material here.

The overhead camera will function like any other television camera with a zoom lens described in Chapter 2. You must locate the zoom, focus, and iris controls and become familiar with their operation. Before you start a class, lay one of your visuals on the copy stand beneath the camera, zoom in as far as the lens will go, then focus the picture. The camera is now focused for the remainder of that class, assuming, of course, you don't hit the wrong control. Also, remember to pay attention to the safe area in the center of the picture, and don't use text or graphics too small to read on the television screen. Now here are some specific uses for this device.

THE ELECTRONIC BLACKBOARD

I have had no problem whatsoever giving up the chalkboard in my ITV classroom. I have always found it difficult to write legibly on it, and I customarily break the chalk, leaving only little nubs which are impossible to grip properly. Plus, I find it rude to turn my back to the students whom I am trying to address. You may feel differently. If so, and you cannot give up the smell, feel, and sounds of chalk on slate, or white dust on your clothes, be sure to mark off the surface into rectangles proportional to the aspect ratio of the television screen, and write within these spaces in letters large enough to be seen at your distant classrooms. Your assistant or aide will need to follow your activity with a camera set up for this purpose.

If you are like me, however, and find it easier to write on a piece of paper than on a chalkboard, then read on. The advantages to the electronic blackboard are its spontaneity,

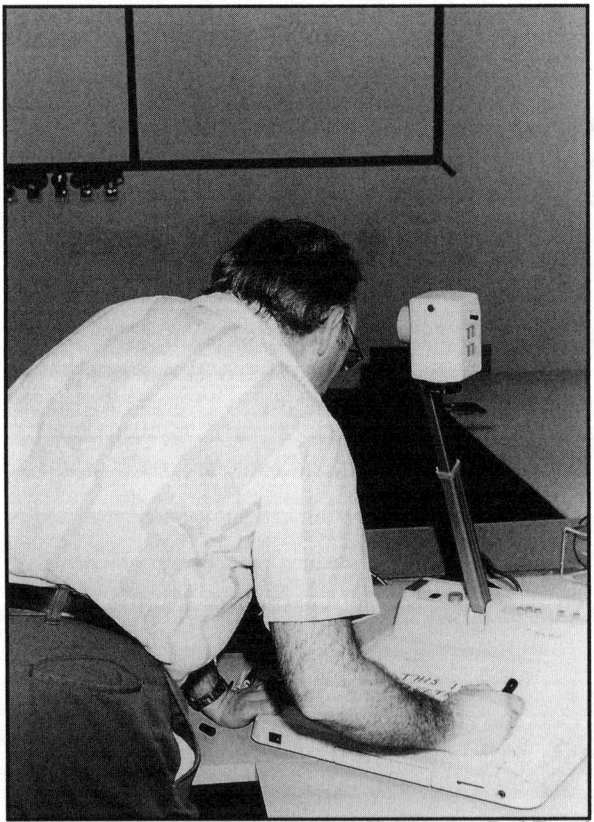

An example of the overhead camera used as an electronic blackboard.

Murray State University, Rod Freed

its simplicity, and its ability to focus students' attention on key points. The disadvantage, as discussed in Chapter 2, is the limited field of view on the television screen. You will not have as much room to write as on a chalkboard, so what you write must be concise. Therefore, you must use this tool for summaries, outlines, key points, and succinct questions. This will involve rethinking some of your lesson plans and breaking the information down into smaller bits which proceed in logical fashion to a conclusion. If you are like me, this will probably be a healthy exercise in itself.

The best tools for the electronic blackboard are a pad of paper and broad-tipped felt markers. We have had pads cut the exact size of the field of view of the overhead camera when the lens is zoomed out to the maximum wide angle, and have marked the copy stand with tape to indicate where the corners of this pad should be placed. Wherever the instructor writes on this pad, the letters will be visible on the television screen. An 8 1/2 x 11 pad turned sideways, to conform to aspect ratio, will suffice. Make sure you use unlined, off-white (tan, gray, or light blue pastel) paper and dark markers.

THE OVERHEAD PROJECTOR

Within certain parameters, artwork created for transparencies will also work with the overhead camera. You may lay the transparency on the pad or duplicate it onto the same off-white paper. Again, the artwork should be composed horizontally, be centered on the page with adequate border area, use large typeface for labels, and be visually simple to understand. Often, the visuals will need to be recomposed for the television format, and some will be totally illegible on the screen. Ironically, I have found that many tranparencies which do not read well on television likewise do not really work on an overhead projector. If in doubt, put up the visual on the screen, then go to the back of the classroom to see if you can make sense of it.

Although you should avoid visuals like a full page of text from a book, or the class schedule from your syllabus, there may be ways you can adapt imperfect visuals by manipulating the television camera and using common sense. If a chart is too complex, or the letters too small, use the zoom lens and move the page around under the camera to focus on a smaller portion of it. Once again, you are breaking down the information into smaller bits and working in a logical sequence. Make sure you communicate with the graphic artist who prepares the visuals so that they will be done with the overhead camera in mind.

Another useful step is to distribute any complex visuals as handouts. The overhead camera then becomes a reference tool supporting active learning. The students can follow each step and even make notes on the handout itself. In addition, each student will get three exposures to the material—from the handout, from the overhead camera, and from the lecture/demonstration/discussion. A further variation of this capability is to distribute handouts with some blanks and have the students collaborate on completing the chart, diagram, text, etc.

THE SLIDE PROJECTOR

There are several ways to transmit slides or filmstrips electronically. One is to transfer the slides to videotape. This method works well if the transfer is done with a telecine, or film chain, set up for this purpose and if the slides or filmstrip include a prerecorded narration. With a videotape, however, it is difficult to pause, backup, or skip ahead as you would do with a carousel projector. And rearranging your slides means either editing the tape or making a new one. A second method is to capture the slides in a computer, store them on a diskette, and play them back through a computer at another time. Although this procedure retains the flexibility of a traditional slide show, the preparation is time-consuming and most floppy disks will store only 4-8 slides, depending on picture resolution. And, of course, many instructors do not have computers with sophisticated graphics capability in their ITV classrooms.

A low-tech compromise is to bring in a slide projector and screen, point the overhead camera at the latter, and run the slide show the way you always have. Many overhead cameras have rotating heads and moveable mounting arms which allow you to aim them elsewhere. The projector screen can be a piece of white posterboard mounted on the wall behind you. Actually, the smaller the projected

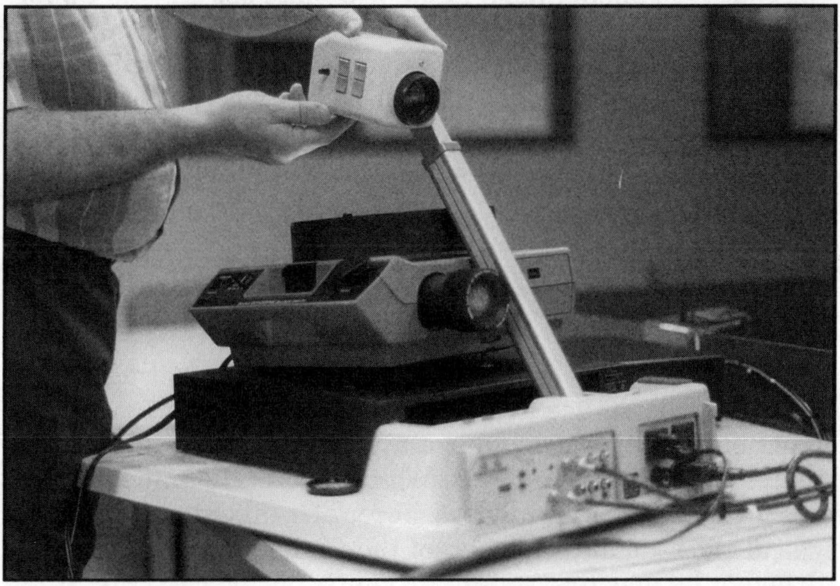

Murray State University, Rod Freed

The overhead camera rotated and positioned to be used with a standard slide projector.

image, the higher both resolution and illumination will be. Make sure that the axis of the projector beam and the focal axis of the camera are parallel so that the picture will be square to the camera. An easy way to accomplish this is to place the projector on the copy stand with the beam directly beneath the camera. You may also have to open up the iris on the camera to compensate for low light level in the projector beam.

For the most part, slides to be used on television should be shot with horizontal composition to conform to aspect ratio. Using the overhead camera as your slide projector, however, gives you zoom capability so that you may use vertical slides and crop the picture where part of it fits on the television screen, an improvement over the traditional slide show.

Your classroom aide should be able to set all this up for you in advance if you remember to communicate your need. Then you can run the slides the same way you always have, using the carousel remote control. Don't worry that your students in the classroom with you cannot see the projector screen. They can watch the pictures on the monitor like your distant students.

THE CLOSEUP CAMERA

Television is a closeup medium. Once you get used to this fact, you can use this property of the medium to instructional advantage. One of my colleagues spends his leisure time reenacting Civil War battles. He made an ITV presentation on this subject wherein he talked about the typical clothing and the food of the Confederate soldier. Although he was dressed in authentic period costume, it was only when he put the sleeve of his jacket under the overhead camera that I could see the texture of the rough brown lindsey-woolsey cloth and feel what wearing that uniform must have been like in the humidity and heat of

Vicksburg or Shiloh. And when he zoomed in to fill the screen with a day's ration, some hardtack and a few grains of parched corn, I could almost taste it.

Now, if he had been in the same room, he could have walked around to let us feel the garment and passed around the foodstuffs, such as they were. This activity is impossible for the distant learner, so you must assist by making visual presentations as up-close and personal as possible. Don't be afraid to zoom in closer, whether it is on an object, a portion of a photograph, a diagram, a chart, or even a single word. It is this capability that makes television work, and it is better to err by framing a subject too tightly than by using a wide angle view.

One common mistake I see is displaying an object by holding it up in front of you for the class to see. Let's say, for example, this is a class in dental hygiene, and you are showing a model of a tooth. When you hold it up in front of you, it looms large in your field of view, but it is only four inches tall to the student fifteen feet away and only a fraction of an inch to the student in the distant classroom. Once you place that model under the overhead camera and zoom in, however, it becomes as large to all your students as it is to you. Which method is more conducive to comprehension and learning?

Think closeups and use the overhead camera for this purpose. If you want some pointers on this production style, watch one of the home shopping (*aka.* tacky jewelry) channels, on cable TV, to see how effective the extreme closeup technique can be.

DEMONSTRATIONS

In a similar vein, you can use the overhead camera to stage demonstrations. You will find that the overhead view keeps your hands out of the picture and provides an intimate, over-the-shoulder perspective for the students. As with any production, you need to rehearse and practice not only the demonstration, but also your ability to follow what you are doing by watching the demonstration on television. This latter ability requires a different hand-to-eye coordination, but you can easily acquire this skill. It is certainly less taxing than most video games.

Again, remember to concentrate on closeup and extreme closeup shots of the objects you are demonstrating. I recently graded a student's instructional video which was supposed to teach me how to assemble a carburetor for a lawnmower engine. His concept of the closeup was to get a tight shot of his hands as he displayed the individual parts and put them together. This was fine if he, or his hands, had been the focus of attention. But the result was a massive pair of hands throughout which dwarfed and partially concealed the carburetor.

There will also be demonstrations for which an overhead shot is awkward or inappropriate. For example, showing a chemical reaction in a beaker would be difficult from this angle, and it is impossible to turn the beaker on its side without spilling the liquid. You can, nevertheless, stage the demonstration on a counter or table top, rotating and repositioning the overhead camera, as described above for slide presentations, to see the action. In this situation, you would want to work from behind the area so that the demonstration is between you and the camera.

For instructors using compressed video, let me add another word of caution. Too much motion or too many quick moves will make the demonstration hard for the distant students to watch. The same is true of zoom moves or repositioning text and graphics in front of the camera. With compressed video, these moves are jerky rather than fluid resulting in a phenomenon I call eyeball whiplash.

A good way to practice is to videotape your demonstration, preferably the version that the distant students will see. Then watch the tape back and ask whether you are learning from the demonstration. You should also watch for correctable problems with motion, gestures, hand placement, and background action which detract from the effect you desire.

COLOR CAPABILITY

If your normal visual aids consist of the chalkboard and an overhead projector, you probably use few if any colors. With the overhead camera, it is simple and desirable to expand your palette.

Color is a marvelous communication tool. Start with one of the outlines you have prepared for the overhead camera. Take several different color highlight markers and divide the subject matter into smaller units, each with its own color code. Or take a list of key concepts, color code the related ones and ask the students to explain the relationships. And don't forget to use different color markers with your electronic blackboard, maybe blue for the basic text and diagrams, and red for emphasis or comment.

Magazines are an excellent source of color photographs and graphics. While some of these may not fit the aspect ratio of the screen and others may be too busy to use, you will be surprised how many visuals you can get from this source. Heretofore, you would need to transfer these visuals to slides, or to a monochrome transparency, or make poor quality copies for handouts. Now all your students can see them in full color with all the impact the photographer or artist intended. Don't neglect less professional visuals, such as snapshots, which are easy to provide and relatively inexpensive.

I wish I had an overhead camera in every classroom where I teach. It is a simple and amazingly effective tool to supplement teaching and learning. Be sure to teach your students how to use it, especially those at the distant sites. It will improve both their interaction and their visual communication skills.

APPENDIX B

USEFUL FORMS AND HANDOUTS

This is the type of material that we typically generate for instructors at the beginning of each semester. Although these items may not fit your situation exactly, they may give you some ideas to create a package of your own.

SAMPLE COVER LETTER

Dear Summer Session ITV Instructors:

The ITV team at Continuing Education and Academic Outreach (CE/AO) would like to take this time to welcome you to the summer session. Enclosed you will find copies of what we will need from you to make this summer session go smoothly. There are handouts on schedules, VCR use, proctors, and a phone list. We hope that this will be as helpful to you as they are to us.

At any time, feel free to call any of the CE/AO staff with questions or problems and we will be glad to work with you. With your cooperation and a little teamwork, this will be a sensational summer.

Sincerely,

Laura Wells
ITV Classroom Coordinator
(762-2160)

Or Call:

Mary Boaz (554-8132)
John Yates (762-4225)
John Hart (762-2052)

CLASS SCHEDULES

To be of a greater help to you this term, it is important that we receive a detailed schedule from you. Please include when you will be traveling to a distant site to teach, give exams, show videotapes, use slides or filmstrips, or have an outside guest. Also, please include what time and where you will be giving your final exam, or when you might need additional assistance. If you can give us a rough schedule at the beginning of the semester and contact us when there are additions or deletions to that schedule, it would be of great advantage to us in aiding you in your classes. If you could get this schedule to Laura Wells, ITV, Collins Center, she will get a copy to each of the sites.

Attached is a couple of examples of how to set up a tentative schedule. You can either run down the whole semester or you can break it down into sections, e.g. travel, tapes, exams, when to give hand-outs, and special needs.

We are here to work as a team and to assist you so that we can have an enjoyable and successful year.

SAMPLE TIMETABLE

SED 431 (ITV)

DATE AND TIME

INSTRUCTOR

6/14/92	ITV/Syllabus/Pretest (handout at distant sites)
6/16/92	Chapter 9 — Learning Disabilities (reading assignment)
6/19/92	Video — How difficult can this be? Part I (Approximate the time)
6/28/92	Site Visit — Teach from MCC
6/30/92	Guest Lecturer — (Name and Which Site)
7/04/92	Exam I (Please have proctor at all other sites.)
7/06/92	Video — Name of the Game (time)
7/10/92	Final Exam — Room 211 at 3:00 (If it is going to be at some other time or place than during your regular class period)

SAMPLE SCHEDULE OF ACTIVITIES

MODERN FILM CULTURE

DATE AND TIME

INSTRUCTOR

SCHEDULE OF VIDEOS, SLIDES, AND FILMSTRIPS

June 2 *Dirty Harry* (videotape; have a copy at one other site for backup)
June 12 *Little Mermaid* (videotape)
June 17 *Dances with Wolves* (videotape)
June 23 The Making of *Robin Hood, Prince of Thieves* (slides)

TRAVEL SCHEDULE

June 3 Deliver from MCC
June 14 Deliver from HCC
June 22 Deliver from PCC

EXAM SCHEDULE

June 9 Exam I (Please have proctor on site)
June 24 Exam II (Please have proctor on site)

GUEST LECTURER

June 8 Oliver Stone at MCC
June 26 Dustin Hoffman at MSU

HANDOUT SCHEDULE

June 3 Chapter 1 packet

June 5 Chapter 2 packet
June 10 Chapter 3 packet
June 20 Chapter 4 packet

(Along with this schedule, you will need to include any special instructions such as how to return exams or any special instructions concerning handouts.)

ITV TECHNICAL SUPPORT AND TROUBLE-SHOOTING

Summer session courses will be taught out of the I & T conference room. A technician will be available at all times either directly in the classroom or across the hall in the office suite. The technician will either be John Hart, Eva Mitchell, or Laura Wells. When you travel to the distant site we will be there to start your class and get things going, and we will be there in case you have any difficulties. If you are at a remote site and are having trouble, the first response would be to notify the MSU room and let the technician help you through the problem. If he or she cannot solve the problem, then he or she is instructed to call either John Hart or Laura Wells.

If for some reason you are not able to

contact the technician at MSU, then wait a few minutes for him or her to contact you. If you feel the problem is serious, try to contact one of the other sites. Listed below are the numbers for all the sites: John's office and home; Laura's office and home; and the ITV beeper.

MSU ROOM # 762-3493
PCC ROOM # 554-8659
MCC ROOM # 825-4379

HCC ROOM # 000-0000

Laura Wells:	Office #	762-2161
	Home #	753-6281
	Beeper #	762-7211 (This is a voice pager; please wait for the tone after you call and then leave your message.)
John Hart:	Office #	762-2051
	Home #	753-5548

Either John or Laura will have the beeper on from the hours of 7:30 a.m. till 8:30 p.m.

Tape use in the ITV room is very easy and convenient. We have a VCR that is used for playing tapes. What we suggest you do is make a copy of the tape to have at one other site as a backup in case something happens and we cannot play the original from the local site. If you will let us know, we will take care of duplication for you.

There are several factors that you need to take into account when playing a tape through the network. The first is how much action is seen on the tape. If there is very much motion, then the students will get very frustrated watching it through the network. If this is the case, we suggest that you make a copy for each site in your class. Also if the tape is longer than 60 minutes, we propose that you make a copy for each site. When you play a tape through the network, it would be to your advantage for us to watch at least five minutes of the tape through the network at least a week in advance of the scheduled showing, so we can anticipate any possible problems.

We can only play 1/2 inch VHS tapes through the network. If your original is a 3/4 inch tape, please bring it to Laura Wells and she will take care of getting it dubbed to 1/2 inch. Please bring any originals that you will want backup copies made of to Laura at least three weeks in advance of when you want to show them and she will get them dubbed for you. All copies must be erased as soon as possible after showing.

All classes are taped and kept on reserve at the circulation desk in the library at each distant site. These are held for two weeks and then are returned to be used again. If you would like to have a copy of a particular class tape, you would need to arrange with Laura Wells and the librarian to have that tape pulled after the two weeks are up and held for you. You will be responsible for making your own copies of these tapes. If for some technical reason a class does not get taped at one site, Laura will make arrangements at the other sites to make a dub and have it sent to the appropriate library.

Examinations

To help you in executing your exams at the other sites, you can either send them through the scheduled courier or you can FAX them to the secured FAX number at each site. To ensure academic honesty, there will be a proctor available at the distant sites. This proctor is instructed to stay with the class during the class period or until the last test is turned in. The proctor will administer the exam per the directions of the instructor. If the proctor suspects dishonesty, the proctor will then call the instructor and follow his/her directive. All exams will be taped unless we are otherwise directed by the instructor.

The exam will be returned to the professor as soon as possible. If there is a separate answer sheet, you can have the proctor fax them to you immediately following the exam. We will return the original to you by mail or courier. We will do our best to get your exam back to you as quickly as possible. Best wishes for a successful summer ITV session. Do not hesitate to contact me at anytime.

Laura Wells

TIPS FROM ITV INSTRUCTORS

- ❐ You can teach as you do in a traditional classroom with appropriate adaptations.
- ❐ Be organized! "Downtime" is magnified on television. Also, know when to be quiet.
- ❐ With practice, the technology should become unnoticed.
- ❐ Student involvement is essential! To improve student involvement, call students by name, use group activities, games, role playing, debates, storytelling, etc.
- ❐ Clarity of presentation and concise instruction are important skills to use.
- ❐ Make the students in the remote sites feel part of the class by maintaining eye contact with the camera. Include the camera in your conversation just as you would another person.
- ❐ Be prepared in advance! Relevant material should be sent ahead to the remote site if it is essential to the lesson.
- ❐ Use clear, well-produced audiovisual aids.
- ❐ Move the camera to a student who is asking a question.
- ❐ Talk slowly and clearly.

- ❐ Pay attention to your appearance.

  Avoid wearing intense colors. Pastels and medium shades look best.

  Avoid small or busy patterns.

  Remember, a microphone will have to be attached to your clothing. A scarf around your neck might muffle the sound.

  Avoid shiny jewelry.

  Makeup should be natural.

- ❐ Smile!

Project INTERACT, Kansas State Board of Education. Used by permission

GUIDELINES FOR INTERACTIVE TELEVISION VISUALS

1. **Keep sentences or phrases short and concise—seven or eight words to a sentence, and three to five sentences to a page.**

Reason: The screen allows for only a small viewing area, and long copy is harder to read.

2. **If much information is necessary to convey the message to the student, please make student handouts.**

Reason: ITV graphics were intended to provide visual variety or an outline to the teaching process. Readability and comprehension are extremely important factors in ITV visual aids. Detailed information does not show well on the screen because of size factors of the font and the screen. Using key words to emphasize points works much better. Handouts relating to your visual aids can be detailed.

3. **Line drawings, if kept simple, are acceptable to make into a visual.**

Reason: Simple graphics work best, where complicated graphics will not produce the desired results.

4. **Type size of the letters in the window should be 18 to 24 points.**

Reason: This type size has proven successful for readability.

5. **Type face (font) should be Helvetica Bold.**

Reason: If copy is kept short, this font is easy to read on the screen.

6. **Paper should be gray text and should be printed horizontally.**

Reason: Picture will be in aspect ratio and contrast will be lower.

GUIDELINES FOR ITV PICTURE COMPOSITION

USE THIS FORMAT AS A GUIDE TO PREPARE

GRAPHICS FOR THE TELEVISION SCREEN

*Print or type all text

*Compose text on the horizontal axis

*Use 10 lines or less

*Limit notes to one concept per page

*Contact the site video manager for the use

of slides, overhead transparencies,

computer output, and videotape playback

ITV UTILIZATION SCHEDULE — FALL 1992

	Monday		Tuesday		Wednesday		Thursday		Friday		Saturday	
8 a.m.		Spanish I		Spanish I		Spanish I		Spanish I		Spanish I	SAT 584 (4 weekends)	
9 a.m.	JRT 491	Honors English		Honors English	JRT 491	Honors English		Honors English		Honors English		
10 a.m.			POL 573				POL 573					
11 a.m.		Physics		Physics		Physics		Physics		Physics		
12 noon												
1 p.m.		French I		French I		French I		French I		French I		
2 p.m.	GSC 101	Political Science	FRE 101	Political Science	GSC 101	Political Science	FRE 101	Political Science	GSC 101	Political Science		
3 p.m.	NUR 605				NUR 404							
4 p.m.			NUR 314				NUR 314					
5 p.m.												
6 p.m.	ENG 530		OSY 615	IDC 101	NUR 601		ADM 663	ENG 101	SAT 584 (4 weekends)			
7 p.m.												
8 p.m.												
9 p.m.												

GLOSSARY

AMPLIFIER: An electronic device which increases the signal in an audio or video circuit.

ASPECT RATIO: The ratio of the height to the width of a field of view. In conventional television, the aspect ratio is three units high to four units wide.

AUDIO: A system of electronic devices which capture the sounds of the world, change them into an electrical signal for control and processing, then change the signal back into sound.

CCD: A charged coupled device, used to convert incoming light to a video signal in a television camera.

CLOSEUP (CU): A television picture with a single small object, such as a face, filling the screen.

CODEC: A coder/decoder device which converts a conventional television signal to a stream of digital bits, so that the signal can be fed down a telephone line, and reconverts the digital signal to television.

CRT: A cathode ray tube; a term used for the video monitor at a computer work station.

DEPTH OF FIELD: The area of a television picture that is sharply in focus.

DISTANCE LEARNING: A learning experience where the student is in a location away from the teacher.

DUB: A copy of a video- or audiotape; to make such a copy.

EDIT: A method of assembling prerecorded bits of audio and video in a new sequence in order to compress time and improve the impact of the material on the audience.

FAX: An electronic device which converts a document to an electric signal and sends it down a telephone line, then reproduces a facsimile of the document at a distant location.

FEEDBACK: An electric signal which re-enters a circuit creating noticeable distortion in the sound or picture, such as when a microphone picks up the output of a loudspeaker; a valuable communication mechanism among human beings evaluating distance learning technology.

FOCAL LENGTH: The distance from the optical center of the lens, focussed at infinity, to the target area on the camera pickup tube or CCD.

LAVALIERE MICROPHONE: A microphone that hangs around the neck or can be clipped to a tie, collar, lapel, etc.

LONG SHOT (LS): A television picture with a wide area, such as an entire classroom, visible.

MEDIUM SHOT (MS): A television picture with most of an object, such as a person from the waist up, visible.

MONITOR LOUDSPEAKER: A device that converts the audio signal back into sound so that we may listen.

MULTIMEDIA: A style of presentation using a combination of various visual and sound media, including audiotape, videotape, film, slides, computer graphics, etc.

NOISE: Any undesirable sound; in an audio or video circuit, any distortion that interferes with the signal.

PC: Personal computer.

SAFE AREA: The area in the center of the television screen, comprising approximately two-thirds of the total screen area, which is most likely to be seen by the audience.

TALKING HEAD: A rather dull camera shot of a human being, usually framed head-and-shoulders, who talks endlessly to the camera.

TELECONFERENCE: An interactive group conference, with widely dispersed participants, that uses some form of electronic media to communicate between the various sites.

TELEVISION CAMERA: A photographic device consisting of a lens, a viewfinder, and electronic circuitry, that converts light to an electric signal and routes it to another device, such as a VCR or monitor.

VCR: A video cassette recorder; a ubiquitous and seemingly simple electronic gadget that no one seems to know how to operate.

VHS: A technical standard for videotape used by most consumer level VCRs.

VIDEO: A system of electronic devices that captures the sights of the world, changes them into electrical signals for control and processing, then changes the signals back to pictures.

VIDEO MONITOR: A television "set" that receives audio and video input signals and converts them to sounds and pictures.

VIDEOCONFERENCE: A teleconference using live video originating from at least one site.

VIDEODISC: A disc that can store audio, video, and data. Most videodisc players use an optical laser to scan the surface of the disc and can easily be controlled by a personal computer for instant access to recorded material.

VIDEOTAPE: A variety of magnetic recording tape, often packaged in cassettes, designed to record, store, and playback audio and video signals.

VOICE-ACTIVATED MICROPHONE: A microphone designed to open the audio circuit whenever it senses sound, typically the human voice, at a predetermined volume.

ZOOM LENS: A camera lens with variable focal length.

SELECTED BIBLIOGRAPHY

Acker, Stephen R. and Steven R. Levitt (1987). Designing videoconference facilities for improved eye contact. *Journal of Broadcasting and Electronic Media, 31*(2), 181-191.

Alten, Stanley (1990). *Audio in media,* 3rd ed. Belmont, CA: Wadsworth Publishing.

Bates, A.W. (1983). *Overview of the use of media in distance education.* (ERIC Document ED 298 948)

Bates, A.W. (1986). *Media in distance education.* (ERIC Document ED 298 958)

Bensusan, H. Guy (1991). Teaching on interactive instructional television: An introduction. Unpublished manuscript, Instructional Television Services, Northern Arizona University.

Breyer, Richard and Peter Moller (1988). *Making television programs.* Prospect Heights, IL: Waveland Press.

Brinkley, Robert, Gary Pavlecho, and Nadeen Thompson (1991). Designing and producing courseware for distance learning instruction in higher education: A nine-step, four-element team approach. *Tech Trends, 36*(1), 50-54.

Chute, Alan (1990). Presentation at the AT&T National Teletraining Center, Cincinnati. April, author's notes.

Cyrs, Thomas E. and Frank A. Smith (1990). *Teleclass teaching: A resource guide,* 2nd ed. Las Cruces, NM: Center for Educational Development, College of Human and Community Services, New Mexico State University.

Cyrs, Thomas E. and Frank A. Smith (1991). Designing interactive study guides with word pictures for teleclass teaching. *Tech Trends, 36* (1), 37-39.

Dede, Christopher J. (1990). The evolution of distance learning: Technology-mediated interactive learning. *Journal of Research on Computing in Education, 22,* Spring, 247-264.

Dempsey, Ellen (1990). First word. *Omni*, April, 8.

Descry, Don E. (1992). On the air in rural Minnesota: Taking education to the students. *Educational Horizons, 70*(2), 84-87.

Dillon, Connie L., Sharon Confessore, and Chere Gibson (1992). Interaction in interactive satellite teleconferencing: Can it be increased? *Journal of Interactive Television, 1*, 43-54.

Distance education through telecommunications: A review of lessons learned (1989). A Special Report of the American Assembly of Collegiate Schools of Business.

Gilder, George (1992). *Life after television: The coming transformation of media and American life.* New York: W.W. Norton.

Gunther, Robert J. (1989). The interactive television classroom. *Educational Facility Planner, 13*(5), 20.

Haynes, Kathleen J.M. and Robert Swisher (1992). Student and faculty attitudes toward compressed video. *Journal of Interactive Television, 1,* 55-64.

Hill, F. (1989). *Tomorrow's learning environment.* Alexandria, VA: National School Boards Association.

Hilliard, R.I. (1985). *Television and adult education.* Cambridge: Schenkman Publishing.

Hosie, P.J. (1987). Television production considerations for satellite delivered ETV. *Journal of Educational Television, 13*(1), 43-56.

Interactive Television Teaching (1989). St. Paul: Minnesota State Department of Education.

Kitchen, Karen and Will Kitchen (1988). *Two-way interactive television for distance learning: A primer.* Alexandria, VA: National School Boards Association.

Koontz, F.R. (1989). Critical barriers to the adoption of instructional television in higher education. *Educational Technology*, April, 45-48.

Lee, Paul (1992). Innovative uses of existing technologies and delivery systems: Four case studies. *Educational Technology, 32*(4), 56-57.

Lochte, Robert H., Viola Miller, and Mary Boaz (1991). An evaluation design for technology utilization in multidisciplinary distance learning projects. *Proceedings of the 9th Annual Conference on Nontraditional and Interdisciplinary Programs.* Fairfax, VA: George Mason University, 251-256.

Lochte, Robert H. (1991a). A forum on distance learning technology. *Humanities and Technology Association Review, 10,* Fall, 1-3.

Lochte, Robert H. (1991b). Evaluating distance education. *Humanities and Technology Association Review, 10,* Fall, 25-33.

Lochte, Robert H. (1992). College student attitudes about interactive television: An interdisciplinary study. *Journal of Interactive Television,1*, 65-74.

Linking for learning: A new course for education (1990). Office of Technology Assessment, Congress of the United States.

Massoumian, Bijan (1989). Successful teaching via two-way interactive television. *Tech Trends*, 34(2), 16-19.

Miller, Viola P. (1991). Distance learning alternative instructional strategies. *Humanities and Technology Association Review, 10*, Fall, 4-11.

Millerson, Gerald (1985). *The technique of television production*, 11th ed. Boston: Focal Press.

Ostendorf, Virginia A. (1991). *The two-way video classroom.* Littleton, CO: Virginia A. Ostendorf.

Patton, Michael Quinn (1981). *Creative evaluation.* Beverly Hills: Sage.

Patton, Michael Quinn (1982). *Practical evaluation.* Beverly Hills: Sage.

Price, Michael A. (1990). Designing instructional television viewing rooms for adult learners. Paper presented at the 6th Annual Conference on Distance Teaching/Learning, Madison, WI.

Project interact: Interactive television communication skills training manual (1992). Topeka: Kansas State Board of Education.

Rifkind, Lawrence J. (1992). Immediacy as a predictor of teacher effectiveness in the instructional classroom. *Journal of Interactive Television, 1*, 31-42.

Rogers, Everett (1983). *Diffusion of innovations,* 3rd ed. New York: Free Press.

Rogers, Everett (1986). *Communication technologies: The new media in society*. New York: Free Press.

Rugg, William J. (1991). Videoconferencing and teleconferencing in education. Paper presented at the 21st Annual Meeting of the Popular Culture Association, San Antonio, TX.

Scott, Greg (1991). E-mail's expanding role in education. *T.H.E. Journal, 18*(11), 54-56.

Stockdill, Stacey H. and Diane L. Morehouse (1992). Case study evaluations of three technology projects: Keys to technology selection and project success. *Educational Technology, 32*(3), 57-58.

Sullivan, Joseph F. (1992). Pupils take to interactive TV class. *New York Times,* February 4, A11.

Tykwinski, Joseph R. and Russell C. Poulin (1991). A practical guide to teleconferencing and distance education. Bismarck, ND: North Dakota University System.

Williams, Fred and Victoria Williams (1991). Amo, amas ...amazing! Interactive television comes of age. *Gannett Center Journal*, Winter, 75-84.

Willis, Barry (1989). Distance-delivered instruction: Making it work. *Educational Technology, 29*(10), 46-47.

INDEX

ABOUT THE AUTHOR

Robert H. Lochte is an assistant professor in the Department of Journalism and Radio-Televison at Murray State University, Murray, Kentucky, and serves as director of the television studios on campus. For the past two years, he has been a consultant to the Division of Continuing Education and Academic Outreach at Murray State involved in creating an interactive television network. Specifically, he has offered workshops to train new instructors, participated in facilities design, and conducted evaluation research.

Born in Nashville, Tennessee, he has an undergraduate degree in English literature from Bowdoin College, a master's degree in broadcasting from Columbia College, and a Ph.D. in mass communication research from the University of Tennessee, Knoxville. Prior to his present career as an educator, Dr. Lochte spent more than two decades in the radio and television business as an owner-operator, program director, producer, director, operations manager, and early morning country music DJ. He has worked in both commercial and educational broadcasting, as well as in cable television. Aside from interactive television, his research interests include direct broadcast satellites, management and marketing in the broadcasting industry, and early wireless technology and inventors.